Mathematics
An Appreciation

Houghton Mifflin Company • Boston

Atlanta • Dallas • Geneva, Illinois
Hopewell, New Jersey • Palo Alto • London

Mathematics
An Appreciation

Michael Bernkopf

Pace University

Library of Congress Catalog Card Number: 74-11710

ISBN: 0-395-18583-1

Contents

For Jeanne,

My goodly company.

To the Instructor

THE TEXTBOOK, and the course you can give from it, has as its primary target the liberal arts student. Generally, such a student will have only a modest high school background in mathematics and little interest in the subject, and is being subjected to a terminal course in mathematics to satisfy a degree requirement. A secondary target is the elementary education major who needs to learn more about mathematics.

The course is meant to be *about* mathematics rather than how to do mathematics. Techniques have been deemphasized and the focus has been placed, instead, upon the ideas of mathematics. Hopefully, this will minimize the study of techniques of mathematics, which are forgotten by the students the day after the final examination. The closest analogue to this course is the music appreciation course, where a student is introduced to some of the ideas of music, but no one expects him to be able to sit down and perform on the basis of it.

The book is *not* meant to be self-contained; a certain amount of basic high-school algebra and geometry is assumed. This is particularly true in matters of notation, where it is expected that a student can read that ab is *a times b,* for example.

For the most part, new strange and technical vocabulary have been kept to an absolute minimum. Students whose orientation is not scientific are wary of notation and special words. Furthermore, these are just other things which have to be learned (or memorized), so that they get in the way of what we are trying to do. Why should we, for example, bother the student with "open sentence" when the probability is greater than $1 - \varepsilon$ that he will never see or hear the term again as long as he lives, nor will it really help him understand the logic involved.

Also, and in the same vein, it is *expected that English is read and understood*. This means that a certain amount of precision has been lost, but this is more than made up by an increase in readability. Thus we talk about the sum of two angles, rather than the sum of the measure of those angles, and we are quite willing to use "the negative of a" for "the additive inverse of a."

Pursuing this reasoning, we have chosen the topics that are essentially familiar or at least stem from familiar situations. No attempt has been made to follow current mathematical fashion, and to include such esoterica (to the student, at any rate) as group and ring theory, etc. The ordered-pair definitions for integers and rational numbers have been ignored as being unsuitable for these audiences. My experience has been that the best way to hold the student's attention is to stay close to his intuition.

My intent has been to provide sufficient material for a one-semester course. The heart of the book, in my opinion, is contained in Parts 1 and 2. Except for the last chapter, each chapter in Part 3 is practically self-contained and may be taken up in any desired order if you exercise a little care. Please use the instructors' guide. There is a great deal more in it than just the answers to the exercises.

Finally, proceed at a relaxed pace. Since it is a terminal course, there is no reason to hurry through in order to keep to a syllabus. Let your class create its own meaningful diversions; allow time for interesting side excursions. Have fun, enjoy; I have, I still do, and I have taught this course often over the years.

To the Student

You ARE ABOUT to embark on a totally new learning experience; you are going to learn *about mathematics*, rather than *how to do* mathematics. The courses which have given rise to your text have been variously nicknamed "Math for Poets" or "Math for people who will never-ever-take-another-math-course,"—which will give you some idea.

This does not mean you will not be expected to think; on the contrary, there are many ideas which will take quite a bit of mental exercise before you can see your way through them. After all, no idea which is totally transparent at first encounter can be worth very much!

Carry the course around in your head with you. Think about it in some of your odd moments; I have found buses and trains good for this. It is much more pleasant to think about mathematics, and a lot more profitable, than to stare off into space; besides, it makes the trip pass more quickly.

Most of all, give the course a chance. Mathematics may not have been your dish of tea before, but this is something new. If you let yourself, you can learn something, and who knows, you might even like it!

Acknowledgments

THE IDEAS EMBODIED in an elementary text are never the work of a single person. They can only be the end result of the author's training, of his reading, of his teaching, of his students, and of a multitude of conversations over lunch, across cold drinks, or sitting around in bull sessions. Yet certain individuals stand out, and I would like to acknowledge my debt to just a few.

Above all, the inspiration for much of the material contained here comes from Prof. Morris Kline of New York University. When I was new in teaching, he gave generously of his time and we have had many interesting and profitable discussions over the years.

Herb Gross of MIT has encouraged me through some bad moments.

My colleagues at Pace University have been most helpful. Some have read early versions of some chapters and were kind enough to provide me with their comments. In particular, my chairman and one of my oldest friends in mathematics, Lou Quintas, provided very concrete support by tailoring teaching schedules, courses, etc., to my requirements.

Lastly, thanks to John C. Adams of Dartmouth College, Super Prof, who got me into teaching in the first place.

It goes without saying that any flaws herein are entirely my own responsibility.

Michael Bernkopf
Pace University
New York, N.Y.

Arithmetic

Monstrat aus nume[?] que[?] virtus possit habere.
Explico p[er]numeru[?] que[?] sit p[ro]portio rerum

Part 1

Mathematics
What Is It?

PART ONE For the past twelve years—give or take a few—you have spent a substantial portion of your time studying one form of mathematics or another. Have you really understood what algebra is, or what geometry is all about, or even why arithmetic works the way it does? We are about to answer these and other questions you may have had about mathematics. We will know that we have been successful if you find many more questions to ask.

Please approach the exercises at the end of the section with an open mind. Many are quite different from any of the problems in mathematics you have seen before; some are designed to serve as the basis for further discussions and therefore don't even have a unique correct answer! You may even find you enjoy doing them! We hope so.

Chapter 1

The Nature of Mathematics

1.1 Introduction

The study of mathematics has been going on for many years—thousands at least. Yet, there have never been more than a handful of "professional" mathematicians, even in the broadest sense, and only a few more informed amateurs. But, in spite of this, the numbers of students who have become involved, if not interested, in mathematics have consistently increased over the years.

Why?

Do we study mathematics to train the mind to think logically? Perhaps, but the training isn't likely to be successful even if the drive to think logically were a consuming ambition of a large segment of our student population. As we shall see, the only kind of logic permitted in mathematics is of little use outside of it. If mathematical logic were useful as a tool in ordinary situations, we might expect to find some uniformity of thought among mathematicians. This is clearly not the case as anyone who is acquainted with even as many as two mathematicians knows.

Then do we study mathematics because the end result is useful? The amount of mathematics most of us will ever use in day-to-day situations is limited. Arithmetic is enough to solve any problem ordinarily encountered in a personal life.

In professional life, it may be another story. The business man and the social scientist will need to know enough about statistics to be able at least to talk to the statistician about the meaning of the results the specialist has obtained for him, how best to use the information, and particularly where the pitfalls in its interpretation lie. Of course, the engineer and the computer programmer will need a host of specialized techniques. Some of these specialists might even be numbered among the professional mathematicians. Similarly, the natural scientist is interested in mathematics.

Nevertheless, large numbers of well-educated people will never need to apply any mathematics beyond simple arithmetic. Why then should they study mathematics at a higher level? Quite simply, to achieve a better understanding of the world in which they find themselves. For better or for worse, we live in a society which is saturated with mathematics. Almost every piece of factual description we use is put before us in mathematical form. We move at so many miles per hour, and in a straight line, or around a curve; we weigh so much; we owe so many dollars. On a more sophisticated level we talk about hardness numbers, viscosity indices, and coefficients of friction, not to mention Dow–Jones averages and the cost-of-living index. Furthermore, we tend to think in mathematical terms; not just in terms of numbers but also in geometric terms. We speak of streets meeting at right angles, circular staircases (which are actually helical), rectangular or square pictures, and so on and on. About the only common descriptive data we don't transmit mathematically are colors, odors, textures, and subjective matters of taste.

This mathematical way of describing things has become so much a part of us that we don't think much about it, but in the span of history it is relatively new. The man of the Middle Ages would not have cared how much he weighed; he would merely have referred to himself as fat or thin or whatever. The primitive tribesman would not have cared how fast, in absolute terms, he was walking; he would only be troubled if he could not keep up with his group or elated if he could outpace it. The very idea of quantitative descriptions of physical phenomena was unknown to Aristotle—the greatest of classical Greek scientists.

For this reason our study of mathematics will be a cultural one. We will see how mathematics was introduced into our society in such a way that it has penetrated every corner of it. Also, we will examine the relationship between actual events and the mathematics we use to describe these events, that is, we will examine this language we call mathematics to see to what extent it is useful, and to what extent it is full of pitfalls.

1.2 Concepts and Abstractions

Before we can begin studying mathematics, we should have a clear idea of what mathematics is. For some, mathematics is only arithmetic; for others, perhaps more sophisticated, it is the study of numbers and their relationships, i.e., algebra. Of course, mathematics is partly this, but there is much more; Euclidean geometry—the geometry you probably studied in high school—is very much a part of mathematics, yet is not included in the above characterization. Probability and statistics are also branches of mathematics which can go beyond algebra, as does calculus. Then there are the combinations, such as analytical geometry, which uses algebraic techniques to solve geometric problems, and which is often used to investigate "spaces" having more than three dimensions. The list is long, and continues to grow.

What we need then, in order to get a proper definition of the collection of subjects we call mathematics, is to find a set of ideas which will encompass all of the above topics (as well as many others which we can't even begin to talk about). On the other hand, we don't want a definition which is so broad that it would include too many ideas, which might be better left, say, to physics or philosophy. To achieve this we shall need some tools.

We can start with the idea of a concept. A *concept* is the recognition that several disparate objects share a common property. Thus, if we look out to sea at a portion of the horizon or at the edge of a well-formed tree trunk silhouetted against the winter sky, or at the path made by a raindrop as it falls, we recognize that all of these objects have a certain property we call "straightness." If we consider a *collection* of elephants consisting of a baby and its mother, or a *collection* of hands consisting only of our own, we recognize that each of these has a certain property we can describe as "two-ness." In each case we have conceptualized by the recognition that even though the collections were different, they had something in common.

This "something in common" can now be considered as an object itself, and such an object created out of the process of conceptualization is called an *abstraction*. In the first example, we had the property of straightness; the abstraction created from this property is the straight line. From the other property we create the concept of the number "two."

Loosely speaking, we can consider *two* to be the noun created from the thought contained in the common adjective of the "two hands" or "two elephants," although the basic idea is more extensive, since it is also contained implicitly in the expression "a left and a right thumb." Notice that neither of these properties has any real existence, except within the mind of man.

Again, a circle is an abstraction. No circles exist in nature. Man has created the circle in his imagination out of all of the almost-circular objects he has ever seen. The Greeks, who probably invented the circle as we now know it, conceptualized it from the cross sections of felled trees, from patterns made in still water by a dropped stone, and from the horizon line as seen from a boat far out in the Mediterranean.

The geometric point is yet another abstraction. No one has ever seen a point; in order for an object to be perceptible, it must have a definite volume. Yet a point is often characterized as being something "without length or depth or breadth." We can only guess at the objects which gave rise to this abstraction, possibly grains of sand, or possibly the process of giving precise meaning to the idea of location.

It is also possible to create new abstractions from existing ones; we might call these second- or higher-degree abstractions. A pentagon has no common natural counterpart; it has been created out of straight lines, held together with the additional abstractions of angles and equality of length. All numbers are abstractions, as we have seen by example; but the fractions were created out of the counting numbers, and the irrational numbers (such as $\sqrt{2}$) were created from the fractions. The latter type of number has no obvious set of antecedents in nature at all; yet it has proved to be most useful, as we shall see.

You will have noted that all of the abstractions we have discussed so far have been outgrowths of counting and of describing the space in which we find ourselves. These are generally considered to be mathematical concepts, and so they are. However, these are not the only ones; as mathematics advances, so does the list of abstractions which come under its purview. Today the list is so long as to be almost limitless.

Yet not all abstractions are mathematical. The idea of the cost of living is an abstraction; the ego and the id of Freudian psychology are abstractions; the notion of class as used by the sociologist is an abstraction, as is the concept of motion in a vacuum. In fact, every social and natural science has its own set of abstractions. One of the things that distinguishes mathematics from the other disciplines is that mathematics deals *only* with abstractions. That is to say, nothing which is not an abstraction can be considered to be a mathematical object. But this is not yet enough to entirely define mathematics; we have the objects; we must now discuss how we deal with them.

EXERCISES 1.2

1. Discover some abstractions of your own. Don't cheat by thinking of the abstraction first and then finding the background objects, but try to do it by considering the real things first.

2. From what natural objects could the sphere have been abstracted? The cone? The cylinder?[1]

3. Even large natural or counting numbers can be considered second-degree abstractions. Show this. (*Suggestion:* Is it possible to precisely visualize exactly 100 objects?)

4. Extend the list of nonmathematical abstractions. Try to determine the realities which gave birth to them.

1.3 Reasoning

There are several methods which man uses to extend the boundaries of his knowledge. First among these is certainly experience—we all learn by doing. Fortunately for mankind, we need not start from scratch at each new generation, for man, uniquely among all animals, has been able to transmit the experience of each succeeding generation indirectly through the medium of the spoken and written word.

But there are many "facts" which we cannot acquire through experience. Only an astronaut has any *direct* knowledge that the earth is a sphere, or that it spins on its own axis. No one has ever seen an atom,

---

[1]If you are totally unfamiliar with the terms used in this exercise, or in any that follow, you should look them up. I suggest you start by consulting a child's encyclopedia in order to get a general idea, and then go to a more adult reference for particulars.

let alone any of its component protons or neutrons or electrons, yet we accept these things as having real existence.

How do we know of them? We have used our powers of reasoning to give us the information. That is, we have used reason to extend our knowledge beyond what it is possible for us to perceive directly with our own senses.

There are basically three kinds of reasoning that we use: induction, deduction, and analogy. Of these, only induction and deduction will be of interest to us, and they represent opposite ways of doing things. *Inductive reasoning* draws general conclusions from particular examples, while *deductive reasoning* draws conclusions about particular examples from general principles.

Inductive reasoning is often called the scientific method, since one of its most prominent uses is the discovery of laws of nature. For example, the theory of our solar system which insists that the earth and the other planets rotate about the sun in regular orbits has been formulated because it is the *best available explanation* of myriads of direct observations. That is, it is a broadly encompassing law which has been drawn from many different experiences. We do not have a proof of its correctness; we are only certain that it fits many facts we do know about, that it is extremely useful for projecting into the future, and that there are no known facts which contradict it.

Yet this method is not without its pitfalls. There is an example due to John Conant, former president of Harvard, which illustrates the difficulties. Observe that if I drink enough scotch and soda I will get drunk; also if I drink enough rye and soda, or vodka and soda, or brandy and soda, or gin and soda, and so on. The inductive conclusion we can draw from these experiences is clearcut; it is the soda which makes me drunk! Of course, a refinement of the experiment will produce a contradictory fact, but until such a refinement is produced, the theory of the inebriating qualities of soda water is perfectly sound as a scientific theory.

To sum up, we observe that, in the inductive process, the argument flows from a series of special cases or particular examples or observations towards a wide inclusive statement which goes beyond the immediate. We could diagram induction like this:

$$\text{Particular} \xrightarrow{\text{Induction}} \text{General}$$

Deductive reasoning, on the other hand, proceeds in quite the opposite direction. That is, properties or relationships about *specific objects* are *deduced* on the basis of what we know about certain *general principles* which we can show govern these objects.

Perhaps the simplest form of deductive reasoning is known as the *syllogism*. It is simple, its structure is readily visible, and we will use

CHAPTER 1 / THE NATURE OF MATHEMATICS

it to illustrate many other types of deduction, some of which we shall use as we go along further.

We will start with a most famous syllogism:

All men are mortal.
Socrates is a man.
Therefore, Socrates is a mortal.

Notice that we have started with a general principle concerning man and mortality. We might draw a picture of that law like this:

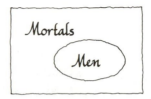

The diagram shows that the set of objects called "men" is entirely included in the collection of objects called "mortals."

The next statement tells us that the object we are interested in is included in the set of men; we could diagram this fact as follows:

If we then put the two diagrams together, we get a picture like this:

This leads *inescapably* to the conclusion of our syllogism, namely, that Socrates is an element of the set of mortals.

We can see in this argument the essential features of any deductive scheme. There are the *hypotheses*, that is, the input of the argument, in this case "All men are mortal" and "Socrates is a man." These generate the *conclusion* which is its output, "Socrates is mortal." The hypotheses will usually include the general principles on which the argument will be based. In addition, there will be statements showing that the objects of concern to the argument satisfy those principles— perhaps under special specified conditions. An argument will be called *valid* if it always proceeds from the more general to the more specific, with no reversals of logical direction. Thus, in order to have a valid deductive argument, it must be inescapable that the general laws we are using are applicable to the particular objects of our conclusion. For example, suppose our syllogism were changed to read

Some men are mortal.
Socrates is a man.
Therefore, Socrates is mortal.

Notice that our first picture is now no longer a good representation of the *new* first sentence. It should be changed to look like this, to allow for all the possibilities inherent in the statement:

The next picture remains the same, but when we come to putting them together we cannot say, on the *basis of the argument alone,* whether

or

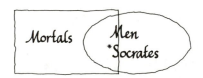

is better as a representation of the argument. The last of the figures would give us our conclusion, but the next-to-last would not. Which are we to take? Since there is no guidance *from the argument,* we don't know. Since we can escape from the conclusion, we must therefore say that the argument is not valid in the deductive sense. *Notice that an invalid argument can have a true conclusion.*

Let's look at a second example.

All spaniels are black.
Mitzi is black.
Therefore, Mitzi is a spaniel.

The picture of the argument looks like this: for the first line,

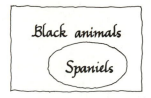

but when we come to draw the one for the second line we are lost. It could be either

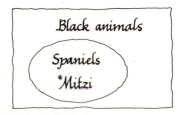

or, just as correctly on the basis of the second sentence:

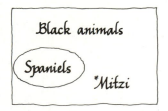

Since there is a perfectly good choice, we are not *forced* into the conclusion, and once again we have an invalid argument. *Notice that this time we have an invalid argument with false hypotheses* (not all spaniels are black) *and a false conclusion* (Mitzi is a mixed breed without a drop of spaniel blood).

Next consider a third syllogism:

All right triangles are equilateral.
A 3–4–5 triangle is a right triangle.
Therefore, a 3–4–5 triangle is equilateral.

Is this a valid deductive argument? Absurd! After all, the sides of a 3–4–5 triangle are all different lengths, so how can they all be equal? Before we leap to the conclusion that the argument is invalid, perhaps we should examine it a bit more carefully. From the first sentence we get a picture like this:

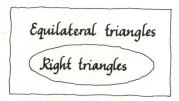

and from the second sentence

Putting them together, we are driven to accept a picture which looks like

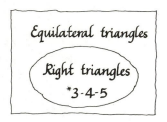

which is the correct picture to give us the conclusion of the argument. Also, notice that there isn't any way we could reasonably represent the hypotheses of the argument which would fail to lead us to the conclusion. Thus we are reluctantly pushed to the point where we must agree that the argument is indeed valid! But how can this be when the conclusion is obviously false!

Clearly, the only place left where we can look for the trouble is in the hypotheses, and of course the difficulty is in the first sentence. To say that all right triangles are equilateral is sheer nonsense! Notice, however, that *in spite of the false hypothesis and false conclusion, the style of the argument is nevertheless still valid.* Thus we see that the validity (or invalidity) of an argument depends upon its form, rather than upon the particular ingredients which go into it.

Now let's look at a fourth example.

> *All oil paintings are great paintings.*
> *Rembrandt painted in oils.*
> *Therefore Rembrandt's paintings are great paintings.*

Our sketches look like this:

and this:

and, finally, this:

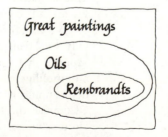

We can see that this is a valid deductive argument with a true conclusion. But one of the hypotheses is false! Then what we have produced is a *valid deductive argument in which a false hypothesis has led to a perfectly good conclusion.*

If we look closely at the last two examples we can see that the deductive validity of an argument will not, *by itself*, guarantee a true conclusion, nor will false hypotheses necessarily lead to a false conclusion, even though a valid argument is used. This tells us that we can never look at the *conclusion alone* to test the validity of an argument or to discover the truth of hypotheses.

Why then do we bother with deductive argumentation? Because, when it is properly combined with true hypotheses, it is infallible. A valid deductive argument in which *all* the hypotheses are true will *always lead to a true conclusion.* This comes about as a result of the form of proceeding from the general law to the special case. In this way, we start with a large or general truth, and at each step we decrease truthfully only the *scope* of the previous statement. That is, what we have at every stage is merely a special case of a larger truth, and so each of the intermediate steps, down to the conclusion itself, must also be true. Notice, that in order to insure truth the direction of argumentation *can never be reversed* to embrace concepts of broader generality than those used in the *previous* step. The argument must always step forward in the order of *decreasing generality* towards its final conclusion.

The infallibility of the deductive argument has a useful auxiliary property. Suppose we know that the conclusion of a certain valid argument is false. In this case, we would then know that *at least one* of the hypotheses going to make up the argument would have to be false.

You may have noticed that we have not attempted to define the term "truth." As we shall see, it is never possible to define every word we use. In any case, the proper definition of such terms as "truth" and "reality" we will have to leave to the philosophers. We will, however, assume that all of us have an idea what they stand for, and even that we are all in pretty good agreement about their meanings.

We will, furthermore, agree that all of the terms we use can assume only two truth values, T or F (for true or false). That is to say, we suppose that, given any statement, there is a mechanism—generally our intelligence bolstered by experience and deductive powers—for determining whether or not a given statement is either true or false. Of course, the truth value of a statement may fluctuate according to circumstances, but *at any one time* only one of these values may be assumed.

What this means is that we must eliminate from our discussions all undecidable or as yet undecided questions and statements, since for such statements the necessary mechanism does not exist. For example, the statement "There are infinitely many prime numbers" is known to be true, while "There are infinitely many prime-number pairs" is as yet undecided. (A prime pair is a pair of *consecutive odd numbers* which are both prime, such as 11 and 13, or 29 and 31.) No one has been able to prove that the latter is true or show that it is false, even though we believe it to be one or the other. (We will discuss proofs later on.) An example of a statement known to be false is, "It is possible to trisect any angle whatsoever using only compass and straightedge." A statement like "This man is good" is undecidable, too, since it calls for a subjective opinion.

EXERCISES 1.3

1. Experiment can be considered as controlled experience. Comment.

2. The gradual disappearance of a ship as it sails away is proof that the earth is round. Comment.

3. The process of conceptualizing and forming abstractions can be considered a form of inductive reasoning. Comment.

4. Conceive your own inductive theory which you know is false, but which can be considered "reasonable" on the basis of some experiments. List the theory and the experiments.

5. An inductive law is said to be *verified* if we can show that a new fact, which was not one used in the discovery of the law, is explained by it. What can you say about a law if there are a large number of verifications? What if there are contradictions? What if there are both?

6. When we mentioned the theory of the solar system, we noted that it is useful for projecting into the future. Is this projection inductive or deductive reasoning? Explain.

7. The word *some* is used differently in mathematics than in ordinary English. In mathematics *some* means the same thing as *there is at least one*. Notice that if "all of something" is a true statement, then that means that "some of the same something" is also true provided that the condition "something" is satisfied by at least one object.
   (a) Write a sentence which means the same thing as "Not all dogs

are black" using the expression "not black." (*Hint: Some* is useful here.)

    (b) Do the same for "It is not true that some dogs are green" using "not green."

    (c) Construct your own examples like those of (a) and (b) and see if you can discover a principle about the use of negations with *all* and *some*. What reasoning process have you used here?

8. Test the following for validity. Comment on the truth of the hypotheses and conclusions.

    (a) *All freshmen are smart.*
        All undergraduates are smart.
        Therefore, all freshmen are undergraduates.

    (b) The same as (a), except that the conclusion is changed to:
        Therefore, all undergraduates are freshmen.

    (c) *Maurice is a Frenchman.*
        All Frenchmen are European.
        Therefore, Maurice is European.

    (d) *The Driver car rides poorly.*
        My car rides well.
        Therefore, my car is not a Driver.

    (e) *All odd numbers are divisible by 3.*
        17 is an odd number.
        Therefore, 17 is divisible by 3.

    (f) *All odd numbers are divisible by 3.*
        17 is divisible by 3.
        Therefore, 17 is an odd number.

9. Supply a valid deductive conclusion for the following:

    (a) *All even numbers are divisible by 2.*
        7 is an even number.

    (b) *All fractions can be put in lowest terms.*
        In the fraction p/q, both p and q are even.

10. If you have done Exercise 7, test the following for validity.

    (a) *Not all dresses are black.*
        This object is a dress.
        Therefore, it is not black.

    (b) *It is false to say that some dresses are black.*
        This object is a dress.
        Therefore, it is not black.

    (c) *This dress is green.*
        Therefore, not all dresses are black.

11. Deduce the "auxiliary property" from the main property of deductive arguments.

1.4 The Definition of Mathematics

We are now ready to define mathematics, and to see where this definition will lead us.

Mathematics is the study of abstractions and their relationships in which the only technique of reasoning that may be used to confirm any relationship between one abstraction and another is deductive reasoning. Thus, by taking measurements in many different triangles, we might convince ourselves that the sum of all the interior angles in any one triangle is 180°, but until we have constructed a valid deductive argument from which we can draw this conclusion it cannot be taken as a mathematical fact.

The matter of using only deductive arguments bears further examination. Recall that every deductive argument must flow from general principles to specific examples. This simple fact requires us to have a set of most general principles in order to have a foundation for our deductions. These most general principles are called *axioms* or *postulates*, and they stand at the very head of the deductive chain.

Following from the axioms are the *theorems*, which become a part of mathematics only when they have been *proved* from the axioms. A *proof* is a valid argument based on the axioms, that is, it is a deductive chain of statements which use the axioms as their most general principles. The conclusion of the proof is the theorem. Of course it is not necessary to go back explicitly to the axioms for every proof. A proof may rely on a previously established theorem, which in turn may itself depend upon a previous theorem, and so on, but ultimately it must be possible to trace a proof back to the axioms upon which it depends. From this we see that the axioms cannot be proved. Since they are *most* general, there is nothing *more* general to serve as the hypothesis of a deductive argument which would lead to an axiom for its conclusion.

A similar situation arises when we come to abstractions. We are forced to use words to communicate with each other, and in order to know precisely which abstractions we are dealing with, we must define the words we are using to describe them. But to define a word, we must use other words, and so on. This means that we cannot define everything without, perhaps indirectly using the word we wish to define in the definition itself. For example, a *straight line* is the path of shortest *distance* between two points on a plane, while *distance* is a measurement taken along a *straight line*. We call such definitions *circular*, and if we try to define everything we are inevitably led to a collection of circular definitions. (Yes, a dictionary is circular!)

We get around this problem in mathematics by assuming that there are certain basic abstractions which are so simple or so fundamental that we will agree that we know what they are. Furthermore, we are at liberty to describe them informally (rather than by a formal definition), and we can use axioms to indicate how they interact with each other. These fundamental abstractions are called *undefined notions* and occupy the same position with regard to definitions that the axioms occupy with regard to proofs. That is, every formal definition uses as its raw material the undefined notions, directly or indirectly.

For example, among the undefined notions in geometry are *point* and the *line*. Anyone who has tried to define these fundamental objects has usually run into the difficulty that he has been forced to use abstractions of a high order of sophistication. Yet all of us have a pretty clear idea of what is involved and the description of a point as "that which has neither breadth nor depth" is really superfluous. One of the axioms of geometry indicating how points and lines interact is the familiar "Every two points determines a unique line." We can then define (formally) a line segment to be "that portion of the line between two points." We can then go on and define a triangle to be the line segments determined by three points, and so on.

To sum up, we will take mathematics to be the study of abstractions where we will allow only the use of deductive reasoning from general principles called axioms. The abstractions with which we are to deal have all been defined from undefined notions. We observe that there is no single set of axioms which apply to all of mathematics. Rather, there is a system of axioms for each branch of mathematics. For example, algebra and geometry each have their own sets of axioms which are quite distinct and have a minimum of overlapping. Each branch of mathematics has this same basic structure.

EXERCISES 1.4

1. If someone told you he had proved the existence of God, what should be the first question you should ask him?

2. When it was thought that the world was flat, a common child's question was "What holds it up?" What has this question to do with axioms and undefined notions?

3. Look up the word *point* in a dictionary. Then look up some key words in the definition, and so on, until you come to a circularity. Confine your attention to the mathematical meaning.

4. Repeat Exercise 3 for *line*.

5. Why is the definition "A point is that which has neither depth nor breadth" flawed? (*Hint:* There are two difficulties, one of concept and one of circularity.)

6. Write out your own definition of a straight line and then criticize it.

7. There is a very crucial undefined notion definition of line segment given in the text. Try to spot it. Can you avoid it?

8. Is the text definition of *triangle* complete? If so, is there anything unexpected about it? If not complete it. (*Warning:* Both answers could be correct, depending on what you are willing to accept or avoid as triangles.)

1.5 Applying Mathematics

Suppose I take four dimes out of my pocket. I could then reason about them as follows: "Four dimes times ten cents gives forty cents." But I have a friend with me who looks at the same collection of dimes and reasons, "Four dimes times ten cents gives forty dimes."

> *Question:* Mathematically speaking, who is correct?
> *Answer:* Both!

The correctness of the answer lies in the fact that the only mathematics involved is the multiplication of four by ten and producing the answer of forty; both have done this much correctly. Recall that mathematics deals only with abstractions, and a dime is certainly real enough, so dealing with dimes is not a mathematical operation. What we are encountering here is the crux of applications; because of the abstract nature of mathematics, some "translation" is necessary if we wish to apply mathematics to real situations. To put things differently, *no problem which is not purely mathematical can be solved by the use of mathematics alone.*

Let us analyze the "dime" problem at greater length, since, in spite of its simplicity, it embodies most of the basic techniques of applied mathematics. In this problem we start with several real objects which we wish to find out more about, using mathematics. To do this, we select mathematical abstractions which best approximate the real objects—in this case, the simple numbers 4 and 10. We also select an operation on the abstractions which will be the best for the type of answer we are looking for—in this case, multiplication. Notice *that we have not yet performed any mathematics;* we have only *translated the problem* into mathematics. Now we multiply 4 × 10 and get 40, which is an abstract answer. Therefore, we must translate this answer out of mathematics in order to get a meaningful answer—in this case, forty cents. But my friend went through exactly the same reasoning up to the final point of translating the result. That is, only his *interpretation* of the answer was different; his mathematics was exactly the same as mine. His difficulty was not a weakness in math, but perhaps a poor background in finance.

We can diagram the entire process of applying mathematics in Figure 1.1.

Figure 1.1

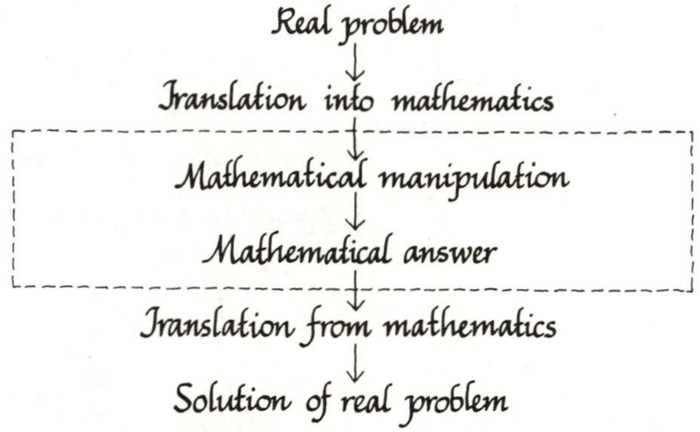

Real problem
↓
Translation into mathematics
↓
Mathematical manipulation
↓
Mathematical answer
↓
Translation from mathematics
↓
Solution of real problem

Note that, in this diagram, the only mathematics which takes place occurs inside the box of dotted lines. This mathematics may be simple, as in our dime problem, or it may be extremely complicated, as in the case of a computer simulation of a manned moon-shot. However, the basic diagram in no way depends on the kind or complexity of the problem we are trying to solve.

Very often the hardest part of a problem comes in translating it into mathematics. If we are only counting, there is no difficulty. We pick the ordinary natural numbers as our abstractions; they were created for just that purpose. But suppose we are working with something more complicated. For example, we might be interested in a problem concerning our earth. What mathematical abstraction could we use? This would depend entirely on the type of problem. If we wanted to look at only a very small piece (say we were making a map of a city), we could assume that the piece was flat; but if we were navigating from New York to London, it would be better to assume that the earth is a sphere. If we were putting a satellite into orbit, we might need to recognize that the diameter through the earth's poles is about 25 miles shorter than through the equator; hence the earth is not a perfect sphere. Finally, if we were considering a problem concerning the earth's own orbit about the sun, it might be sufficient to assume that the earth is a point!

This process of selecting the appropriate mathematical abstraction for applications is called *idealization*, and we have seen that a given object can be idealized in many different ways, depending upon the kind of problem we wish to solve. Furthermore, it is the idealization which is put through the mathematical processes, so that we always work with abstractions.

The basis for many so-called paradoxes stems from the attempt to apply mathematical processes to real objects; we know this is impossible.

We can sum up the entire relationship between abstractions, concepts, and idealizations in Figure 1.2.

We see that it is the process of idealization which closes the ring and brings mathematics to bear on reality.

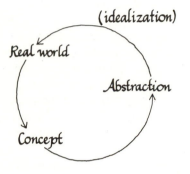

Figure 1.2

EXERCISES 1.5

1. A pessimist looks at a partially filled glass of milk and says "My glass is half empty." If he has two such glasses, is it true that he has no milk? After all, $\frac{1}{2}$ empty plus $\frac{1}{2}$ empty equals 1 empty, equals empty. Comment.

2. A train with two railroad cars has 50 people in each. We multiply 2 cars by 50 people and get 100 people. The local model-railroad club has 50 members, each of whom brings 2 cars to the monthly meeting. We multiply 2 cars by 50 members and get 100 cars. How can we multiply cars by people and in one case the answer is people, and in the other it is cars?

3. Would it be reasonable for a weather forecaster to assume that the earth is a homogeneous and smooth sphere? Explain.

4. Would it be appropriate to idealize the moon as a point when considering eclipses? When considering its effect on the tides?

5. Suppose I have a dime and I square it. On the one hand, 10 × 10 equals 100, so a dime squared is a dollar. On the other, .10 × .10 equals .01 so a dime squared is a penny. Does this mean that a penny is equal to a dollar? Explain.

1.6 The Role of Inductive Reasoning

We have seen that, in mathematics, only deductive reasoning is permitted, yet inductive reasoning cannot be entirely rejected. It has two extremely important auxiliary roles.

The first of these is in the *discovery* of mathematical facts called theorems. It is rare indeed that the deductive processes lead to discovery; deduction is very much an after-the-fact phenomenon. How, then, are theorems discovered? Essentially, a mathematician will work out and go over many particular cases of a theorem which interests him, examining them for appropriate hypotheses, and trying to determine whether there are assumptions hidden in his examples which he is not aware of.

For example, he might measure the interior angles of many triangles and determine by induction that the sum of the interior angles is always 180°. But he must be careful not to measure only the angles in right triangles or isosceles triangles; and he should include both scalene and obtuse triangles. In fact, he should examine triangles in as many shapes and sizes as he can think of. Observe that, after he has done all this, he has only an educated guess, not a theorem; an educated guess is sometimes called a *conjecture*. This conjecture does not become a theorem of mathematics until it has been proved deductively, using the axioms of Euclidean geometry.

A great deal of care must be exercised, however, in the use of induction. Consider the series 2, 9, 16, 23, 30. Each entry is obtained from the previous one by adding 7. Therefore the next entry is 37, right? Wrong! The next entry in *my* sequence is 6 because the series happens to be the day of the month Saturdays fall on starting with October 1976.

The other mathematical use of induction is in connection with axioms. As we have seen, axioms are incapable of being proven; their position at the head of the deductive chain prevents it. It used to be thought—by the Greeks initially, and eventually right up to the nineteenth century—that an axiom was a "self-evident truth." Today, this is no longer the case. What is "self-evident"? For that matter, what is "truth"? Philosophers have been struggling for centuries with this problem. Even if we could discover working definitions of these notions, how could we apply them? How universal would they have to be? The difficulties are just too many.

Modern mathematicians take the following point of view: An axiom is a general statement which will be *assumed* as being true. Does this mean that just any statement can be taken as an axiom? No, not even in theory.

In the first place, no axiom can be chosen which *leads to an inconsistency*. That is, we cannot allow a system of axioms which will lead to deductive proofs both of a statement and of its negative. In other words, we cannot have a system in which we can prove the same statement to be both true and false.

In the second place, since our eventual aim is to apply mathematics to the real world, our axioms must correspond to the truth as we believe it to be. (Note that we are using terms like *truth* and *reality* in the sense of the undefined notions of Section 1.4 above.) How, then, do we discover these truths? By the only course left open to us, inductive reasoning from experience.

For example, it is an axiom of algebra that for all numbers x and y, x plus y equals y plus x. Of course, there is no way to prove this within the structure of algebra. Do we know that it is "true"? Not really, since there are infinitely many numbers to test and it is impossible to check them all out. But in the whole of human experience no one has yet, insofar as we know, been able to produce a pair of numbers whose sum is one value when added one way, and another when the order of summation is reversed. Strictly speaking, all we can really assert with confidence is that we *do not know that the axiom is not true*. In other words, the best we can do with our axioms is to verify them inductively.

This has many interesting consequences, but we will explore only one. As we have seen, a valid deductive argument guarantees a true conclusion only if the hypotheses are all true. But we have also seen that in mathematics, we don't really know whether our hypotheses are true or not, although we suspect that in fact they are. This means that we cannot really assert that our results represent truth solely because they have been produced mathematically. In the last analysis, mathematics rests, through the axioms, on the same empirical basis as physics or sociology or medicine.

How, then, are we to test the truth of our results? The best way is to test them against reality. Our system is only as solid as the accuracy with which it describes the real world. Thus we are forced from the traditional view that mathematics confirms reality into the view that reality confirms mathematics.

EXERCISES 1.6

1. The expression $n!$ is a short way of writing $1 \times 2 \times 3 \times \cdots \times n$. Thus $2! = 1 \times 2 = 2$, $3! = 1 \times 2 \times 3 = 6$, $4! = 1 \times 2 \times 3 \times 4 = 24$. Multiply 4! by 5, and 5! by 6. Guess inductively at a theorem. Also, notice that $2! < 5^2$ and $3! < 5^3$ and $4! < 5^4$. (The symbol $<$ is read "is less than.") Can you guess at another theorem? Are both of your theorems correct?

2. Consider the sequence 365, 365, 365, 366, 365, 365, 365, 366, 365, 365, 365, . . . What do you think is the next entry? Is it the only possibility? Explain?

3. We define a fraction to be a number which can be written in the form p/q, where p and q are integers, and q is not zero. We define the product of two fractions $(p/q) \times (r/s)$ to be the fraction pr/qs. Why is it not a good idea to analogously define the sum $(p/q) + (r/s)$ to be

$$\frac{(p + r)}{(q + s)}?$$

4. "If something can be proved mathematically it must be true," is an old saw. Comment on it.

REFERENCES

The references following most chapters provide a partial bibliography of books of particular relevance to that chapter. Consult the bibliography in the back of the book for a more complete listing, and don't forget your library for further reading.

Barker, Stephen F. *Philosophy of Mathematics.* Prentice-Hall, Inc., Englewood Cliffs, N.J., 1964.

Bell, Eric Temple. *The Handmaiden of the Sciences.* Reynal & Hitchcock, Inc., New York, 1937.

Broehm, George A. W., and the editors of *Fortune*. *The New World of Math*. The Dial Press, New York, 1959.

COSRIMS. *The Mathematical Sciences: A Collection of Essays*. The M. I. T. Press, Cambridge, 1969.

Dubnov, Ya. S. *Mistakes in Geometric Proofs*. Translated by A. Henn and O. A. Titelbaum. D. C. Heath and Company, Lexington, Mass., 1963.

Hogben, Lancelot. *Mathematics for the Millions,* 4th ed. W. W. Norton & Company, Inc., New York, 1968.

Kline, Morris. *Mathematics, A Cultural Approach*. Addison-Wesley Publishing Co., Inc., Reading, Mass., 1962.

Kline, Morris. *Mathematics and the Physical World*. Doubleday Publishing Company, Anchor Books, Garden City, N.Y., 1963.

Kline, Morris. *Mathematics in Western Culture*. Oxford University Press, New York, 1953.

McKay, Herbert. *The World of Numbers*. Macmillan Publishing Co., Inc., N.Y., 1946.

Newman, James R. *The World of Mathematics*. 4 vols. Simon & Schuster, Inc., New York, 1956.

Newsom, Carroll V., *Mathematical Discourses: The Heart of Mathematics*. Prentice-Hall, Inc., Englewood Cliffs, N.J., 1964.

Polya, G. *How To Solve It*. Doubleday Publishing Company, Anchor Books, Garden City, New York, 1957.

Renyi, Alfred. *Dialogues on Mathematics*. Holden-Day, Inc., San Francisco, 1967.

Russell, Bertrand. *The Impact of Science on Society*. Columbia University Press, New York, 1953.

Sutton, O. G. *Mathematics in Action*. G. Bell & Sons, Ltd., London, 1958.

Whitehead, Alfred North. *The Interpretation of Science*. Edited by A. H. Johnson. The Bobbs-Merrill Co., Inc., New York, 1961.

Whitehead, Alfred North. *An Introduction to Mathematics*. Oxford University Press, New York, 1958.

Part 2

Old Acquaintances Revisited

PART TWO In this part we are going to take a look at certain particular topics from the mathematics you have looked at before. Obviously, we can't go back over everything—nor would we if we could. *This is not intended to be a review.*

Our aim is to take a fresh look at certain aspects of arithmetic, algebra, and geometry in light of what we now understand mathematics to be, and in light of how it interconnects with reality. In the process we hope we can clarify some things that may have puzzled you. Once we have pointed the way, perhaps you can unravel some private mysteries of your own which we will not have discussed.

The main point is this: *many of the apparently unreasonable formulations with which you struggled are in fact only supremely obvious developments from the real world.*

We will assume that you have met up with a little algebra as well as a little geometry and arithmetic. This will allow us to use symbols from algebra wherever it is convenient, even when we are discussing the other branches (although logic and history tell us that things didn't happen that way).

Chapter 2

Arithmetic

2.1 The Counting Numbers

The basis of all arithmetic, both historically and logically, is the set of numbers 1, 2, 3, 4, . . . These are called the *counting* (or *whole* or *natural*) numbers; and it is not hard to see how these numbers developed from a need for counting (hence one of their names). We shall take them to be the *undefined notions of arithmetic*, along with the usual operations of addition, subtraction, multiplication, and division.

The first thing to observe about these numbers is that the sum and product of counting numbers is again a counting number, a property called *closure*. This seems so obvious it hardly needs commenting upon, but is it? What about subtraction and division? Is it true that the natural numbers are closed—that is, have the property of closure—under these latter two operations? Of course, $12 - 7$ is a natural number, as is $4 \div 2$. But in order for a set of numbers to be closed under an operation, that operation must *always* produce a number in the given set. Now, since $6 - 15$ and $3 \div 7$ are *not* counting numbers, we have shown that the counting numbers are *not closed* under the operations of subtraction and division. Observe that it was necessary to produce only a *single example* in order to show that the property of closure was not satisfied.

The next point to notice about the natural numbers is that the abstraction, that is to say, the number itself, is independent of the strokes or marks which we use to represent it. In some ways we are all familiar with this idea; we know that the abstraction "five" can be represented by strokes ⋕ or by the Arabic-numeral symbol 5, or by the Roman-numeral symbol *V*, or even by the word *five*.

But there is another way of representing numbers with unfamiliar symbols—or more accurately, with familiar symbols arranged in an unfamiliar way. We all know from experience that we can represent any counting number we wish by arranging numbers selected from the collection 0, 1, 2, 3, 4, 5, 6, 7, 8, 9 in an appropriate order. This collection contains the first nine counting numbers, along with a new symbol, zero. It is no accident that it takes only ten different symbols to represent any number we choose—it is a direct result of the fact that we write our numbers using *ten as a base*, which in turn is probably a consequence of the fact that we have ten fingers. That is to say, we write our numbers in such a way that the column in which each digit appears counts appropriate *powers of ten*.

For example, 623 represents the number in which there are 6 ten-squareds (hundreds), 2 tens (that is, tens to the first power) and 3 units (that is, tens to the zeroth power). (Recall that any nonzero number raised to the zeroth power is one.) Using more sophisticated notation, we could write $623 = 6(10)^2 + 2(10)^1 + 3(10)^0$. Again, the number 89,607 could be written in this same form as $8(10)^4 + 9(10)^3 + 6(10)^2 + 0(10)^1 + 7(10)^0$. Notice the importance of zero coefficient of $(10)^1$ as it is used here: a placeholder. Without the zero, there would be no way to indicate that the second-to-last column—the tens column—is empty. Zero used in this fashion as a placeholder is the key to this manner of representing numbers which is called *positional notation*.

This leads us to the thought that perhaps we could write numbers using bases other than ten. Suppose, for example, that, like Mickey Mouse, we were all born with four fingers on each hand. Then we might expect to live in a world which uses base eight instead of base ten. For example, $(613)_8$ (the little eight to the right of the last parenthesis is called a subscript and is used to indicate that we are dealing with base eight rather than ten) can be written in the notation we used before as $6(8)^2 + 1(8)^1 + 3(8)^0 = 6(64) + 8 + 3 = 395$ in the familiar base ten. To take another example, $(1074)_8 = 1(8)^3 + 0(8)^2 + 7(8)^1 + 4(8)^0 = 512 + 0 + 56 + 4 = 572$ (base ten). Again, we see the use of zero as a placeholder to tell us about the empty second column.

Notice that only eight symbols, 0, 1, 2, 3, 4, 5, 6 and 7, are necessary to represent a number in base eight. Furthermore, $(8)_8$ and $(9)_8$ are meaningless; actually $8 = 1(8)^1 + 0(8)^0 = (10)_8$ and $9 = 1(8)^1 + 1(8)^0 = (11)_8$. This is an example of a general rule. Any counting number b, except for the number one, can be selected as a base for representing numbers, and there will be a need for exactly b different symbols. Thus, the smaller the base, the fewer the different kinds of symbols which will be needed. This is balanced by the fact that, in general, a longer string of these symbols is necessary to represent the same number. $(572)_{10}$ needs only a string of length three in base ten but (as it is equal to $(1074)_8$) a string of length four in base eight.

Again, we emphasize that $(572)_{10}$ and $(1074)_8$ are merely different methods of representing the same abstraction. Neither is better or more correct than the other; it is just that we are more used to base ten as compared to other bases.

We can use this idea of different bases to clarify the mystery of the "borrowing" and "carrying" we found necessary when adding or subtracting. An example in base eight will illustrate. Suppose we want to solve the problem $(1074)_8 + (613)_8$. We could arrange our work in the familiar way:

$$(1074)_8$$
$$\underline{(\ 613)_8}$$

being careful to keep appropriate columns lined up in order to be sure we are counting the same things in both cases. We can translate that, using our notation, into:

$$1(8)^3 + 0(8)^2 + 7(8) + 4$$
$$6(8)^2 + 1(8) + 3$$

and, adding column by column, we get an answer

$$1(8)^3 + 6(8)^2 + 8(8) + 7$$

Note that we now have obtained an additional 8^2 from the second column from the right, so we really have $7(8)^2$ and $0(8)$. Rewriting the answer as

$$1(8)^3 + 7(8)^2 + 0(8) + 7$$

or, somewhat more conventionally,

$$(1074)_8$$
$$(\ 613)_8$$
$$(1707)_8$$

Another way of looking at what we have just done is to recall that the symbol for eight in base eight is $(10)_8$. Hence, when we added $7 + 1$ in the second column we got $8 = (10)_8$. We wrote down the last digit (zero) and "carried" the first digit (one) to the next (third) column to indicate that we had more powers of eight than we could handle in the second column. Let's try one in base ten using this same notation. Consider

$$863$$
$$+\ 659$$

which we rewrite as

$$8(10)^2 + 6(10) + 3$$
$$6(10)^2 + 5(10) + 9$$

Adding column by column we get the answer

$$14(10)^2 + 11(10) + 12$$

Taking the right-hand column first, we write $12 = 10 + 2$. Now our answer becomes:

$$14(10)^2 + 11(10) + (10 + 2) = 14(10^2 + (11(10) + 10) + 2$$
$$= 14(10)^2 + 12(10) + 2$$

that is, we have recorded the last digit of the last column, but we have taken the overflow and "carried" it into the second column, since the units column cannot accomodate any number larger than 9. Now the tens column of our answer looks like 12×10, which is larger than the 9×10 which it can hold. Therefore we write 12×10 as

$$(10 + 2) \times 10 = 10 \times 10 + 2 \times 10 = 10^2 + 2 \times 10$$

Replacing this in our answer for the tens column, we have

$$14(10)^2 + (10^2 + 2(10)) + 2$$
$$= 14(10)^2 + 10^2 + 2(10) + 2$$
$$= 15(10)^2 + 2(10) + 2$$

so we have "carried" the excess 10^2 into the lefthand column. Now we turn our attention to this column, which is also overflowing since its capacity is 9×10^2. To correct this we write

$$15(10)^2 = (10 + 5)(10)^2 = 10^3 + 5(10)^2$$

so that our answer now looks like:

$$(10^3 + 5(10)^2) + 2(10) + 2$$
$$= 1(10)^3 + 5(10)^2 + 2(10) + 2$$
$$= (1522)_{10}$$

We can also apply the same principle to subtraction; the only difference is that we reverse the process. Suppose we wish to find $(2074)_8 - (613)_8$. We rewrite as before

$$
\begin{array}{r}
2(8)^3 + 0(8)^2 + 7(8) + 4 \\
(-) \qquad 6(8)^2 + 1(8) + 3 \\
\hline
\end{array}
$$

but now we *subtract* column by column. The last two columns on the right give us no trouble, since we can take 3 from 4 and 1 from 7; but at column two there are difficulties, since we cannot take 6 from zero. To get around this problem we "borrow" one of the 8^3 from the first column and write it as $8(8)^2$, so that our problem now looks like:

$$
\begin{array}{r}
1(8)^3 + 8(8)^2 + 7(8) + 4 \\
(-) \qquad 6(8)^2 + 1(8) + 3 \\
\hline
\end{array}
$$

When we subtract on a column-by-column basis we have

$$1(8)^3 + 2(8) + 6 \times (8) + 1 = (1261)_8$$

for our answer.

Clearly the technique which we used does not depend upon the particular base in which we have written our numbers, in particular the familiar base ten. Therefore, the details are left for you to supply.

EXERCISES 2.1

1. Go back to Exercise 1.6.3. If you have not worked this exercise, do so. How do the results apply to the discussion on closure?

2. Are the integers closed under subtraction? Division? Are the fractions closed under subtraction? Division?

3. Explain why no more than 10 symbols are needed for base-10 representation. Why are at least ten needed?

4. Write the following in the base 10.
$(63)_8$ $\qquad$ $(241)_8$ $\qquad$ $(1002)_8$
$(72013)_8$ $\qquad$ $(923)_8$

5. Write the following in the base 8.
$(196)_{10}$ $\qquad$ $(64)_{10}$ $\qquad$ $(24)_{10}$ $\qquad$ $(320)_{10}$ $\qquad$ $(513)_{10}$

6. For those who know a little French: Does the French method of naming numbers suggest anything about the shoe-wearing habits of the ancient Gauls? (*Hint:* What do they call 80, 73, 92?)

7. The base two has come into prominence in recent years. What are its advantages? Its disadvantages? Why would it be especially useful for computing devices?

8. Explain why the number "one" cannot be chosen as a base for representing numbers.

9. Suppose we encounter a tribe of intelligent three-toed sloths who have developed an arithmetic. What base (or bases) could we reasonably expect them to use to write their numerals? Write out a multiplication and an addition table for whatever base b you picked. Why is it unnecessary to go beyond $(10)_b + (10)_b$ and $(10)_b \times (10)_b$ in your tables?

10. Calculate, following methods of the text.
$(123)_8 + (310)_8$, $\qquad$ $(732)_8 + (27)_8$, $\qquad$ $(641)_8 + (756)_8$,
$(526)_{10} + (194)_{10}$, $\qquad$ $(326)_{10} + (789)_{10}$,
$(430)_{10} + (72)_{10}$.

11. Calculate the following, using the usual "carrying" technique.
$(6)_8 + (12)_8$, $\qquad$ $(36)_8 + (61)_8$, $\qquad$ $(25)_8 + (173)_8$,
$(24)_5 + (131)_5$, $\qquad$ $(1001)_2 + (1011)_2$,
$(12112)_3 + (21221)_3$.

12. Calculate, using methods of the text.
$(14)_8 - (6)_8$, $\qquad$ $(213)_8 - (62)_8$, $\qquad$ $(1064)_8 - (726)_8$.

13. Apply the methods of the text to explain the process of indenting from the right that we use when multiplying multidigit numbers. (*Hint:* Try some two-digit numbers first.)

°14. Use exercise 13 to show how an ordinary adding machine without a multiplication key can be used to multiply 213×574 by pushing the repeat addition key exactly six times.

2.2 The Integers

The next step up in complexity from the natural numbers is the set of integers, . . . , -2, -1, 0, 1, 2, . . . ; that is, the set of numbers containing the counting numbers, zero, and their negatives. Historically speaking, the negative numbers are not very old, not as old as, say, the positive fractions or even some irrationals such as the square roots. Even René Descartes (1596–1650), one of the fathers of modern mathematics, refused to accept the negative numbers as part of the number system.

The integers are formed essentially by taking differences of natural numbers, that is, by extending the idea of subtraction beyond that which is possible using natural numbers alone. Specifically, we can consider the integers an extension of the natural numbers, obtained by permitting the subtraction of any natural number from any other natural number, without regard to size. Thus we can get the integer -3 by subtracting 7 from 4 or 12 from 9 or 1035 from 1032. In particular, we can get zero by subtracting any number from itself.

Zero is an interesting number. We have already seen that it is indispensable as a placeholder, but this is only one of its special properties. Perhaps its most outstanding characteristic is its neutrality under addition and subtraction. That is, if we add zero to or subtract zero from any number, we still have the same number we started with. Mathematicians would indicate this property by writing

"For every number a, a + 0 = a."

We shall study other special properties of zero later on.

Unfortunately, this neutrality of zero under addition has led many students to the erroneous conclusion that zero is nothing. But zero can't be nothing since it is at least a number, with all of the structure of a number. That, of course, makes it an abstraction. What we *can* say is that zero, far from being nothing, is the abstraction of nothing. Perhaps an example will help; in the First National Bank, where I do not maintain an account, my balance is nothing, while at the Merchant's Bank, where my account is located, my balance may well be zero after a period of heavy expenditures.

Zero also leads us very naturally to the concept of the negative of a number. Because of the neutral property of zero, we can ask: Is there one number which we can add to a given number to get an answer of zero? Or, informally, what number *undoes* whatever a given number does under addition? In symbols, given a, is there an x so that $a + x = 0$? Note that this x undoes whatever a does under addition, for if we start with n and add a, we get the new number $n + a$; if we then add x to this new number we get $n + a + x = n + 0 = n$ again. To repeat, *the negative of a number a is that number which, when added to a, gives zero.* We designate the negative of a by $-a$, which is read "negative a" or "minus a" or even "the opposite of a."

Sadly, the terminology here is unfortunate. Students often confuse "the negative of a" with "a is negative" when there is really only a slight connection between the two statements. Actually, $-a$ is positive (greater than zero) just about as often as it is negative (less than zero). For example, what is the negative of -5? It is that number which, when added to -5, gives zero. But 5, when added to -5, gives zero; therefore, 5 is the negative of -5. In fact, from similar reasoning, we can show that the *negative* of *any negative number* is positive. It would be easier or more logical if the language could be modified or changed, but the terms have been around much too long.

There is another mystery of the integers which has long puzzled many students: Why should the *product* of two negative numbers be positive? This appears to violate the pattern established by the positive integers, where the product of two positives is again positive.

There are actually two very sound reasons why the product of two negatives should be positive, one intuitive and one technical.

Let's take the technical reason first. We all know, and accept as reasonable, the fact that $(-a)b = -(ab)$, particularly if a and b are positive. That is, the negative of a, when multiplied by b, is the same as the negative of the product ab. In addition, we can justify this intuitively by observing that if we have five bags of grain, each losing three pounds, we have lost 15 pounds of grain, or $-3 \times 5 = -15$, where we have identified the concept of loss with the negative numbers.

Now suppose, for the moment, that the product of two negative numbers is negative; in symbols, suppose $(-a)(-b) = -(ab)$. Let's try it with some figures. If

$$(-3) \times (-5) = -15$$

but, from our grain-bag example above,

$$-3 \times 5 = -15$$

then

$$-3 \times 5 = -3 \times -5$$

Now, the equals sign tells us that -3×5 and -3×-5 are merely different ways of representing the same number, so when we divide each by the same nonzero number, the answers should also be equal; but after dividing both sides by -3, we have

$$5 = -5$$

which is ridiculous. What has happened here is that we have fed several hypotheses into a valid deductive argument and arrived at a false conclusion. Therefore we know that at least one of the hypotheses is false, and a careful examination shows that it must be the supposition that $(-3) \times (-5) = -15$. Taking $(-3) \times (-5) = 15$ leads to no such contradiction.

The other reason for taking the product of two negatives to be positive is that it corresponds to nature. Let us consider a large tank of gasoline which loses 3 gallons an hour by evaporation. If we multiply the number of hours by (-3) we will get a product which indicates the amount of the change in the quantity of fluid in the tank relative to the present time. Thus, two hours from now the change in the gasoline will be $-3 \times 2 = -6$ gallons, that is, a loss of 6 gallons (as is indicated by the minus sign in the answer). Now it is a reasonable translation of time into mathematics to take the future as positive and the past as negative. Thus, four hours ago translates as -4. What was the state of the gasoline four hours ago? Our experience tells us that there were 12 gallons more in the tank than now. But that is precisely the answer we get if we multiply -4 (the number of hours from now) times -3, that is, $(-4) \times (-3) = 12$, where the positive answer represents an *increase* in the amount of fluid.

Finally, we want to see if we accomplished what we set out to do when we looked at the integers, namely, to get a set closed under subtraction. Now it is clear, from the way that the integers were constructed by subtraction, that the difference of any two *counting numbers* is defined, but what about the difference of any two integers? To explore this question, we must understand the notation $a - b$. What we really mean by $a - b$ is $a + (-b)$, that is, the difference between a and b is the sum of a and the *negative of b*. In this way, we have changed the problem of the closure of integers under subtraction to the question of closure of the integers under addition; or, to put things differently, for the integers, subtraction is just a modification of addition. To completely discuss the matter of the closure of the integers under addition gets a little complicated, but with tools of greater sophistication, the mathematician can indeed show that the integers satisfy the closure property under addition, using as a basis the closure property of the counting numbers under addition.

EXERCISES 2.2

1. We all know that numerically $-a = -1 \times a$. Yet they are entirely different in concept. What is this difference?

2. Invent two of your own illustrations to show that zero is not nothing.

3. Note that from the definition, $-(-a)$ is that number which when added to $-a$ gives zero. Show that $-(-a) = a$.

In the exercises below, you may assume that *zero times anything is zero*. Also, you may need the distributive law, $r(s + t) = rs + rt$.

4. Show that $(-a)b = -(ab)$ by showing that $(-a)b$ is that number which, when added to ab, gives zero. What tacit assumption are you making here?

°5. Use Exercises 3 and 1 to show that $(-a)(-b) = ab$.

6. Referring to the gasoline-tank problem in the text, how is the present time now translated into mathematics? Does this agree with the rest of the problem?

7. Distinguish carefully between the negative numbers and the negative of a number.

8. Would it be possible to consider subtraction as a form of addition if we were dealing with numbers which had no negatives? Explain.

2.3 Rationals and Division

The problem of closure under subtraction which we have just studied has its companion problem; namely, now that we have the integers, can we extend our numbers to the point where we can always divide one integer by another? The utility of such a process is obvious; clearly we will often want to study a part of an entire object, yet we will still want to retain a way of relating the amount under consideration to the whole.

The answer to this closure question is, "Almost!"

To see how we can extend the integers but not quite so that we have closure under division, let's take a closer look at the dividing process. What exactly do we mean by $x = a \div b$ or $x = a/b$? What we are really doing is searching for a number x with the property that, when x is multiplied by b, the answer will be a. Thus, $2 = 6 \div 3$ means precisely the same thing as $2 \times 3 = 6$.

In symbols, we are looking for x so that $bx = a$. Note that division "undoes" multiplication the same way that subtraction "undoes" addition.

If we now confine our attention to the integers alone, we see that it is not always possible to divide one integer by another. For example, $(-3) \div 7 = x$ has no solution *in the integers*, since there is no integer x which will make $x \times 7 = (-3)$ a true statement.

Why, then, can't we employ the same kind of device we used when we extended the counting numbers to the integers? Wouldn't it be possible to define a new collection of numbers to be the results of taking quotients of integers? Yes, we can do this if we exercise a little care, and take a look at a special problem first.

What are we to mean when the divisor is zero? This is a tough nut to crack, and the difficulties stem from the fact that the *product* of zero and any other number is always zero. Now, suppose we try to *divide* an integer a by zero; we write the problem just as we did earlier: $a \div 0 = x$. This of course means that we seek an x so that $x \cdot 0 = a$ is a true statement.

At this point we must separate the problem into two cases, depending upon whether a itself is zero or not. In the first case, we suppose a is not zero; but then there is no possible x which makes $x \cdot 0 = a$ true, since $x \cdot 0 = 0$ no matter what value x assumes.

The other possibility is that a is zero; that is, our problem now looks like $x \cdot 0 = 0$. But in this case we are embarassed by having too many solutions, since, for *every* integer x, the expression $x \cdot 0 = 0$ is true.

In either case we are in trouble. Either there are not enough (actually, no) x's which will fill the bill, or there are too many. The mathematician gets around this difficulty by insisting that it is not an allowable mathematical operation to divide by zero; that is, *division by zero is not defined*.

But then, since division by zero has no definition, couldn't we exercise our prerogative of creating new objects out of old by defining the quotient of a number divided by zero to be, say, a number? Since division by zero is not otherwise defined, there could be no conflict. Let's see what happens if we try; consider the following:

Let $\qquad\qquad a = b;$

then $\qquad\qquad a^2 = ab \qquad\qquad$ Multiplying both sides of the equation by a

and $\qquad a^2 - b^2 = ab - b^2 \qquad$ Subtracting b^2 from both sides

and $\qquad (a - b)(a + b) = b(a - b), \qquad$ Factoring each side and so $\qquad a + b = b;$ $\qquad\qquad$ Dividing both sides by $a - b$

thus, $\qquad\qquad 2b = b$ $\qquad\qquad$ Since $a = b$

or $\qquad\qquad\qquad 2 = 1$ $\qquad\qquad$ Divide both sides by b

which is absurd.

Here we have a valid deductive argument with a conclusion which is obviously false! In fact, if this conclusion were allowed to stand, then every integer would be equal to every other integer, obviously an impossible state of affairs. But what went wrong here?

Recall that when we had a valid deductive argument with a false conclusion, the place to look for the difficulty was in the *hypothesis*. In this case, a careful examination of the argument shows that the only questionable act occurred at step four where we divided by zero in the form of $(a - b)$, for, since $a = b$, $a - b = 0$. Everything else is perfectly allowable. What we have shown is that if division by zero were an allowable mathematical act, then every integer would be equal to every other integer. So when mathematicians say that division by zero is not defined, they are not just ducking a thorny issue; they have very real reasons!

With this in mind, we can take the *rational numbers*, or fractions, to be the numbers formed from *quotients of integers*, provided that the dividing integer is *not zero*. In symbols, a rational number is any number which can be written in the form $a \div b$ or a/b where a is any integer and b is any integer which is not zero. Notice that if the rational number r equals a/b, this is exactly the same as saying that $rb = a$.

Just as zero is the neutral element under addition, the number "one" is the neutral element under multiplication. As we did in the previous section, it is reasonable to ask whether there is a unique number which will produce an answer of one when multiplied by a given number, say a. The answer is yes, unless, of course, a is zero. We call the number the *inverse*—or, if we wish to be explicit, the *multiplicative inverse*—of a, and write it $1/a$ or a^{-1}. Again, the former notation is suggestive, since if $r = 1/a$, then $a \times r = 1$. We also observe that $1/ab$ is $1/a \times 1/b$ since, if we multiply $1/a \times 1/b$ by ab we have

$$ab\left(\frac{1}{a} \times \frac{1}{b}\right) = ba\left(\frac{1}{a} \times \frac{1}{b}\right) = b\left(a \times \frac{1}{a}\right)\left(\frac{1}{b}\right)$$

$$= b \times 1 \times \frac{1}{b} = b \times \frac{1}{b} = 1$$

Thus $1/a \times 1/b$ is that number which, when multiplied by ab, produces an answer of one, or $1/a \times 1/b = 1/(ab)$. It is this fact which now leads us to define the multiplication of two rationals. If $r = a/b$ and $s = c/d$, then we take $r \times s = ac/bd$.

The definition of the sum of two rationals is not so simple. It is tempting to assume that the sum of $r = a/b$ and $s = c/d$ is

$$\frac{(a + d)}{(b + d)}$$

following the example set by the product of r and s.

There are two difficulties with this definition. The first is an aesthetic one. If we used this definition, the sum of two rationals would not always be a rational. It could happen that $b = -d$, in which case the divisor would be zero, something which is not permitted. This, by itself, wouldn't be an insurmountable obstacle; we could just avoid adding two such fractions.

But the second difficulty cannot be surmounted. It gives an incorrect result—not just a different answer, but the *wrong* answer, when it is checked against experience. Certainly, in any system involving fractional parts, we want $\frac{1}{2} + \frac{1}{2} = 1$. But with the proposed definition of addition that we have under consideration,

$$\frac{1}{2} + \frac{1}{2} = \frac{(1 + 1)}{(2 + 2)} = \frac{2}{4} = \frac{1}{2}$$

This result just doesn't agree with our experience! To put things differently, this "obvious" definition of addition *gives the wrong answer!*

In order to get a definition which will give us a correct answer, let us look at some specific problems whose answers we know beforehand from the real world.

If we add $\frac{1}{2}$ to $\frac{1}{2}$, note that we are adding things which are expressed in the same kind of units, namely halves. In this case there are two of them, so we have two halves (expressed by $\frac{2}{2}$) or 1.

Now suppose we wish to add $\frac{1}{4}$ and $\frac{1}{6}$. Here we have things expressed in different kinds of units (fourths and sixths); experience suggests we should convert both fractions to the *same kinds of units*. In this case we pick the easiest, 24ths. We can convert fourths to 24ths by multiplying the divisor, 4, by 6. But recall that one is the neutral element under multiplication. In order to multiply the divisor by six, we must multiply the *entire* rational, $\frac{1}{4}$, by 1 written as $\frac{6}{6}$, so that

$$\frac{1}{4} = 1 \times \frac{1}{4} = \frac{6}{6} \times \frac{1}{4} = \frac{6 \times 1}{6 \times 4} = \frac{6}{24}$$

Similarly, we convert $\frac{1}{6}$ into $\frac{4}{24}$, in this case multiplying by $\frac{4}{4}$. Now we are adding objects expressed in the same units,

$$\frac{1}{4} + \frac{1}{6} = \frac{6}{24} + \frac{4}{24} = \frac{(6+4)}{24}$$

$$= \frac{10}{24} = \frac{2 \times 5}{2 \times 12}$$

$$= \frac{2}{2} \times \frac{5}{12} = 1 \times \frac{5}{12}$$

$$= \frac{5}{12}$$

(The last four steps are necessary only if we wish $\frac{10}{24}$ to be in lowest terms.)

This technique is perfectly general. Suppose we wish to add a/b and c/d. We convert both to the same units, and the most convenient is bdths. We then multiply a/b by one in the form d/d, and multiply c/d by one in the form b/b. Once this is done, we find we are adding ad of the bdths to cb of the bdths; this gives us $(ad + cd)$ of the bdths, which we write as

$$\frac{ad + cb}{bd}$$

We can use the principle of positional notation when it comes to representing fractions, just as we did with the integers. This time we use the columns to represent powers of $1/10$. We put the decimal point in front, in order to warn the reader that we are dealing with a fraction and not an integer.

In this way, when we write .6302 we mean $6(1/10) + 3(1/10)^2 + 0(1/10)^3 + 2(1/10)^4$. Following our definition for adding fractions, this can be written as $6,302/10,000$. Similarly, any terminating decimal, that is, one with only a finite number of terms, can be written as a fraction. But what about nonterminating decimals? Are these fractions? The answer is, only some are, while some are not.

Suppose we consider .333 · · · (the three dots mean "continue on indefinitely exactly as before"); call it N:

$$N = .333 \cdot \cdot \cdot$$

then

$$10N = 3.333 \cdot \cdot \cdot$$

If we subtract the top from the bottom (that is $10N - N$), we get

$$9N = 3 \qquad \text{or} \qquad N = \frac{3}{9} = \frac{1}{3}$$

The key fact about the demonstration above was that $.333 \cdots$ was a *repeating decimal:* in this case, the 3 repeats indefinitely. More generally, any repeating decimal can also be written as a fraction. As an example, consider

$$N = .430143014301 \cdots$$

Then $10{,}000N = 4301.430143014301 \cdots$ Subtracting N from $10{,}000N$, we get

$$9999N = 4301$$

or $N = 4301/9999 = 391/909$.

Here the repeating block was 4301 and the only "trick" in the demonstration was to multiply N by a suitable power of 10, so that when we are ready to subtract, one block appears in exactly the right position so that the "tail" is zero when we subtract. It should be easy to see that this principle can be applied to any repeating decimal.

A repeating decimal does not have to start with its repeating block. Thus $0.763201201201 \cdots$ is a repeating decimal, the only requirement being that *after some point* the repeating block takes over. To show that even this type of decimal can be written as a fraction, let

$$N = .763201201201 \cdots$$

Then

$$1000N = 763.201201201201 \cdots$$

and, after subtracting N from $1000N$ (the upper expression from the lower), we have

$$999N = 762.438$$

Multiplying both sides by 1000 to get whole numbers on both sides, we get

$$999{,}000N = 762{,}438 \qquad \text{or} \qquad N = \frac{762{,}438}{999{,}000}$$

Conversely, we can show that *every fraction* can be written as a repeating decimal. Again, we shall resort to a convincing example rather than a proof. Consider 7/13. If we actually perform the long division we get

$$\begin{array}{r} .538461538\cdots \\ 13\overline{)7.000000\cdots} \end{array}$$

1	65
2	50
3	39
4	110
5	104
6	60
7	52
8	80
9	78
10	20
11	13
12	70
13	65
14	50
15	39
16	110 etc.

There are two things to notice here. The first is that, when dividing by 13, there are at most 12 nonzero remainders. The second is that we are "bringing down" only zeros at every step after the first, and this process continues indefinitely. These two facts mean that at some point (here, line 14) a particular remainder *must* repeat; and that once the repetition has started, the whole process will continue to form the pattern exactly as before. To continue the process from line 16 above, we see that 13 goes into 110 eight times with a remainder of 6 (just as in lines 4, 5, and 6); after bringing down the zero, we find that 13 goes into 60 four times with a remainder of 8 (as in lines 6, 7 and 8); and so on, so that the whole block 538461 is going to show up again, and then again, and again. Thus, we have shown that $7/13 = .538461538461\cdots$

As we have seen so often, the process is perfectly general. In order to show that a/b can be written as either a terminating or a repeating decimal, we actually perform the long division of $a \div b$. From some point on, we will be bringing down only zeros, and there will be no more than $(b-1)$ different nonzero remainders. (If we get a zero remainder the process stops, and we have a terminating decimal.) Thus, we only have to reach the position where we have the same remainder we had earlier, and when we bring down the zero, we will exactly repeat a calculation we made before. Once this has started, we are locked into the repetition process, just as we were in dividing 7 by 13, and we are led to a repeating decimal.

Now we have shown that every repeating decimal is a fraction and that every fraction can be written as a repeating decimal. This means that whenever a number has one property it has the other; that is, the two concepts are mathematically indistinguishable. But certainly not every decimal is repeating. Consider .303003000300003 · · · , where we keep inserting *one more zero* between successive threes at each step. We can now assert that there is no fraction which represents this nonterminating decimal.

We also know there are numbers which cannot be written as fractions. In the next section, we shall show that $\sqrt{2}$ is such a number; and consequently now we know that it is not a repeating decimal either. The numbers which cannot be written as fractions (and hence, cannot be written as repeating decimals) are called *irrational numbers*.

EXERCISES 2.3

1. Show that if $2 = 1$, then $3 = 2$ and therefore, $3 = 1$, and so on, so that $3 = -6$, etc.

2. Explain why $1/0$ is a meaningless set of symbols. Do we improve anything if we write $1/0 = \infty$?

3. Explain the meaning, if any, of $0/0$.

4. What important assumption was used tacitly when we showed that $1/a \times 1/b = 1/ab$?

5. Show that $1^{-1} = 1$. Are there any other integers which are their own multiplicative inverses?

6. When we were looking at our preliminary definition of the addition of rational numbers, why didn't we test the result by adding $13/17$ and $9/7$?

7. There are two and only two things (which are essentially different) which we can do to a number without changing its value. Comment.

8. When we were discussing the conversion of fourths to twenty-fourths in the text, we started off by multiplying the divisor by six. Why didn't we start by adding six?

9. When you first studied the addition of fractions, your teacher probably spent a great deal of time showing you how to find the *least* common denominator. Was this necessary? Explain.

10. Show that $0/a$ is the neutral element for addition for the fractions, no matter what nonzero value a assumes.

11. In the text we discussed the decimal representation of fractions only between zero and one. Why is this no serious limitation?

12. In showing that .333 · · · = $\frac{1}{3}$, where did we use the fact that .333 · · · was nonterminating?

13. Is it true that 1.000 · · · = .999 · · · ? Support your guess with reasons.

14. Is it true that .667 = $\frac{2}{3}$? .66667 = $\frac{2}{3}$? Is any terminating decimal equal to $\frac{2}{3}$?

15. Show that every terminating decimal can be considered as a repeating decimal.

16. Convince yourself that every fraction can be written as a repeating decimal by trying $\frac{5}{7}$ until you have at least *two entire repeating blocks*. If you are still unconvinced, make up some of your own examples and try them.

17. Can $\sqrt{2}$ be written exactly?

°18. The equivalence of fractions and repeating decimals does not depend upon the fact that we use base 10. Try to show this.

19. Show that between every two rationals there is another rational; that is, show that if r and s are rationals, with r bigger than s, there is a rational smaller than r but bigger than s. (*Hint*: Let r and s be two rationals. Look at the number which is the average (mean) of the two. Try with examples first.)

°20. Show that between any two rationals there are more than K rationals, where K is any counting number whatsoever. (*Hint*: Use Exercise 19 and proof by contradiction.)

°21. Show that between any two rationals there is an irrational. (*Hint*: Use the fact that the rationals can be written as repeating decimals while the irrationals cannot.)

°22. Show that between two rationals there are more than K irrationals, where K is any counting number whatsoever.

2.4 Irrational Numbers

The existence of numbers which cannot be written as fractions has already been demonstrated through the medium of decimal representation. But the question arises, what is the use of such numbers? Why do we need yet another new class of numbers?

As usual, we will see that when we try to perform a new kind of operation, in this case root extraction, the existing numbers cannot handle the new situation. Suppose we wish to find $x = \sqrt{2}$; that is, we are looking for a positive number which, when squared, gives 2. Clearly, no counting number will do, because $1^2 = 1$, so 1 is too small, and $2^2 = 4$, so 2 is too big. Our next attempt might be 1.5, but $(1.5)^2 = 2.25$, so 1.5 is too big. The number $(1.4)^2$ is less than two, as is $(1.41)^2$, while $(1.42)^2$ is greater than two; so $\sqrt{2}$ lies between 1.41 and 1.42. If we continue in this way, we find that $\sqrt{2}$ is between 1.414 and 1.415, and between 1.4142 and 1.4143, and so on. We notice that we keep trapping $\sqrt{2}$ between two fractions (recall that every terminating decimal is a fraction), which are getting closer together at each step; but we would find that no matter how long we persevere, we never could pin the elusive $\sqrt{2}$ down as a terminating decimal. Furthermore, we might also notice that there was no apparent repeating block of decimals either.

From this, we might suspect that there was no fraction which was equal to $\sqrt{2}$; and in fact, this suspicion is correct. Of course, we haven't *proved* that there is no such fraction; we have only an inductive guess. In order to prove this conjecture, we must begin from a different direction. Instead of looking at the decimal representation, we examine the problem by considering a fraction as the quotient of two integers.

For this we need some facts about the fractions and integers. First, every fraction can be put into lowest terms; that is, if we have a fraction a/b, we may assume that a and b have no common factors. (This, surprisingly, is difficult to prove.) Second, we will need to know that the square of an even integer is even,[1] and the square of an odd integer is odd. These last two facts together tell us that if an even integer is a perfect square,[2] its square root must be even. (To see this, note that its square root must be odd or even; but if the root were odd, the square would be odd, which it isn't, by hypothesis. Therefore, the root can't be odd, and hence must be even.)

We now are ready for our proof. Suppose there *were* a fraction $a/b = \sqrt{2}$.

Then we may assume a and b have no common factors.	First assumption above, that every fraction can be put in lowest terms.

---

[1]An integer n is even if $n = 2k$ for some integer k.

[2]A perfect square is an integer which is the square of some integer, such as 1, 4, 9, 16, 25, 36, . . .

Now $a^2/b^2 = 2,$	Square both sides.
$a^2 = 2b^2,$	Multiply both sides by b^2.
a^2 is even;	a^2 is written as $2b^2$; that is, a^2 is 2 times an integer.
a is even;	Since a^2 is even
$a = 2K$	(See definition of even.)
$a^2 = 4K^2$	Square both sides.
$4K^2 = 2b^2$	Substitute for a^2 from above.
$2K^2 = b^2$	Divide both sides by 2.
b is even.	Since b^2 is even

And here we have an absurdity. On the one hand a and b have no common factors, but on the other hand, the assumption that $a/b = \sqrt{2}$ led us to the fact that they are both even (that is, are both divisible by two).

Here we have a valid deductive argument with a ridiculous consequence. Where did we go wrong? A careful examination will show that the only difficulty comes from the initial assumption, that there existed integers a and b with $a/b = \sqrt{2}$. Therefore, this initial assumption must be false; that is, we are forced to the conclusion that $\sqrt{2}$ is irrational.

In general we can state that $\sqrt{a/b}$ is irrational unless both a and b are perfect squares after the fraction is put in lowest terms; or, more generally, $\sqrt[k]{a/b}$ is irrational unless a and b are both perfect k-powers after the fraction has been reduced to lowest terms.

We should pause for a moment to clear up what is often a source of confusion about square roots. Recall that, if a mathematical symbol appears with no sign in front of it, by agreement the plus sign is supposed to be present. Thus $\sqrt{7}$ is the *positive* square root of seven, while $-\sqrt{7}$ is the negative root. Now, if we ask what are *all* of the numbers whose square equals seven, they are $\sqrt{7}$ and $-\sqrt{7}$. We represent *both* of these numbers by the symbol $\pm\sqrt{7}$. Notice that symbols like $\sqrt{-7}$, or $\sqrt{a}$ for any negative a, are meaningless at this stage.

When operating with square roots, we know, from those values of r and s that we can test, that $\sqrt{r \cdot s} = \sqrt{r} \cdot \sqrt{s}$, where r and s are rationals (we are including integers now as rationals since they can be written in the form $a/1$). For example, we see that $\sqrt{4 \times 9} = \sqrt{36} = 6$ on the one hand, while $\sqrt{4} \times \sqrt{9} = 2 \times 3 = 6$ on the other.

But when we come to addition, there is another story. It is definitely *not* correct to say that $\sqrt{r + s} = \sqrt{r} + \sqrt{s}$. Let's try some examples. If $r = s = 1$, we have $\sqrt{r + s} = \sqrt{1 + 1} = \sqrt{2}$ while

$\sqrt{r} + \sqrt{s} = 1 + 1 = 2$. Certainly $\sqrt{2}$ is not 2, since $\sqrt{2}$ isn't even rational! Even if both sides turn out to be rational, we don't get the right answer:

$$\sqrt{25 + 144} = \sqrt{169} = 13$$

but

$$\sqrt{25} + \sqrt{144} = 5 + 12 = 17$$

Root extraction of rationals, together with certain sums, gives us the means of building up a large set of irrationals, but not all irrational numbers can be formed in this way. Those that can be are called the *algebraic irrationals*; but there is a whole class of irrational numbers called the *transcendentals*. These, although more numerous than the algebraics, do not appear commonly; and, in fact, the only transcendental number which is generally known outside of the sciences is pi (π), the ratio of the circumference of a circle to its diameter.

EXERCISES 2.4

1. Show that $\sqrt{2}$ is between 1.41421 and 1.41422.

2. Show that the square of an even integer is even. (*Hint*: Recall an integer x is even if and only if it can be written in the form $x = 2k$ for some integer k.)

3. Show that the square of an odd integer is odd.

4. $\sqrt{2}$ was probably the first number known (by the Greeks) to be irrational. It arises very naturally in geometry. Give a geometric construction.

5. Prove that $\sqrt{3}$ is irrational. Be careful to state all your assumptions, and prove as many as possible.

6. The rationals are closed under addition and multiplication. Does it follow that the irrationals are, too? Illustrate your answer by a proof or by examples.

7. Show that the sum of a rational and an irrational must be irrational. What about their product?

8. When we checked out the formula $\sqrt{r \cdot s} = \sqrt{r} \cdot \sqrt{s}$, why didn't we take $r = 13$ and $s = 7$, say?

9. Only under very special conditions on r and s will $\sqrt{r + s} = \sqrt{r} + \sqrt{s}$. Find these conditions.

*10. Try to prove that $\sqrt{4}$ is irrational, following the pattern of proof used in the text to show the irrationality of $\sqrt{2}$. Since $\sqrt{4} = 2$, which is rational, what goes wrong in your proof? Why *must* something go wrong? Explain.

11. Verify, by example, that if r is positive, then

$$\sqrt{\frac{1}{r}} = \frac{1}{\sqrt{r}}$$

Use facts demonstrated in this section to show that if s is also positive, then

$$\sqrt{\frac{s}{r}} = \frac{\sqrt{s}}{\sqrt{r}}$$

REFERENCES

Aaboe, Asgar. *Episodes from the Early History of Mathematics*. Random House, Inc., New York, 1964.

Beckmann, Peter. *A History of π*, 2d ed. Golem Press, Boulder, Colo., 1971.

Cajori, Florian. *A History of Elementary Mathematics*. Macmillan Publishing Co., Inc., New York, 1924.

Cajori, Florian. *A History of Mathematical Notations*. 2 vols. Open Court Publishing Company, La Salle, Ill., 1928.

Dantzig, Tobias. *Number, the Language of Science*. Macmillan Publishing Co., Inc., New York, 1954.

Gamow, George. *One, Two, Three . . . Infinity*, rev. ed. The Viking Press, Inc., New York, 1961.

Gow, James. *A Short History of Greek Mathematics*. 1884. Revised reprint. Chelsea Publishing Co., Inc., New York, 1968.

Heath, Thomas. *A History of Greek Mathematics*. Oxford University Press, Inc., New York, 1960, Vols. I, II.

Heath, Thomas. *A Manual of Greek Mathematics*. 1931. Reprint. Dover Publications, Inc., New York, 1963.

Hofmann, Joseph E. *The History of Mathematics*. Translated by F. Gaynor and H. O. Midonich. Philosophical Library, Inc., New York, 1957.

Hogben, Lancelot. *Mathematics for the Millions*, 4th ed. W. W. Norton & Company, Inc., New York, 1968.

Huntley, H. E. *The Divine Proportion: A Study in Mathematical Beauty.* Dover Publications, Inc., New York, 1970.

Kline, Morris. *Mathematics and the Physical World.* Doubleday Publishing Company, Anchor Books, Garden City, N.Y., 1963.

Kushyar ibn Labban. *Principles of Hindu Reckoning.* Translated by M. Levey and M. Petruck. University of Wisconsin Press, Madison, 1965.

Newman, James R. *The World of Mathematics.* 4 vols. Simon & Schuster, Inc., New York, 1956.

Peter, Rozsa. *Playing with Infinity: Mathematics for Everyman.* Translated by Z. P. Dienes, Simon & Schuster, Inc., New York, 1964.

Pullan, J. M. *The History of the Abacus.* Praeger Publishers, Inc., New York, 1968.

Reid, Constance. *From Zero to Infinity.* Thomas Y. Crowell Company, New York, 1960.

Van Der Waerden, B. L. *Science Awakening.* Translated by A. Dresden. Oxford University Press, New York, 1961.

Willerding, Margaret. *Mathematical Concepts: A Historical Approach.* Prindle, Weber & Schmidt, Inc., Boston, 1967.

Chapter 3

Algebra

3.1 Introduction

Once again we are going to go over ground which should be familiar to you; but hopefully the approach will place old ideas into new settings, so that you can look at them from a fresh standpoint. We remind you that this is not going to be a review of algebra, nor will you be asked to learn any new techniques. What we are going to do is discuss the algebra you already know from our position that mathematics is simply an extension of reality.

A word of caution: The term *algebra*, as we use it here, is the basic elementary algebra studied in the high schools. The mathematician uses the word *algebra* in a much broader sense; he will use it to mean the study of almost any abstract system, together with the operations (akin to addition and multiplication) defined on that system. One example that you may already have encountered is Boolean algebra; and in a later chapter we will discuss what might be called the algebra of sets and the algebra of statements.

Furthermore, in what follows, we shall freely use all of the rules of dealing with equality. Most of these rules stem from our assumption that the equal sign indicates that a particular number (or collection of numbers) is represented in two different ways. We list some of them for your convenience.

$a = a$	Any object is equal to itself.
$a = b$ whenever $b = a$, and vice versa.	The order of writing down an equality is immaterial.
If $a = b$ and $b = c$, then $a = c$.	Things equal to the same things are equal to each other.
If $a = b$, then $a + c = b + c$.	The same number may be added to both sides of an equality without destroying the equality.
If $a = b$, then $ac = bc$.	The same number may multiply both sides of an equality without destroying it.
If $a = b$ and if $c \neq 0$, then $a/c = b/c$.	An equality may be divided on both sides by the same nonzero number without destroying it.

3.2 Algebra: Super Arithmetic

What is algebra? There are usually two answers to this question; first, algebra is the language of science and mathematics; and second, algebra is generalized arithmetic. Of course, both of these are correct, but they both need a considerable amount of amplification in order to make sense.

Essentially algebra is arithmetic, only more so. All of those operations we can do with numbers, and *only those*, are precisely the operations of algebra. The only real difference between algebra and arithmetic is that in arithmetic we are operating with particular numbers, while in algebra we operate with symbols which *represent* numbers.

Let us be more specific. We start, as in any mathematical structure, with our undefined terms. In this case our undefined terms will be a set of symbols (usually letters of the alphabet), all of which represent numbers, the numbers themselves, and all of the operations on the numbers. With this in mind, it makes sense to add or multiply symbols alone, such as $x + y$, or $a \cdot b$, or in combination with numbers, such as $9 \cdot w$ (which we more often write as $9w$), or $3 - x$. Once again we note that we are dealing essentially with numbers, so that anything which cannot be done numerically cannot be done algebraically.

Since algebra is an outgrowth of arithmetic, it is not surprising that we turn to arithmetic for its axioms. These axioms are necessary because we are dealing with new objects, symbols which represent numbers, not just the numbers themselves. The axioms tell how the symbols interact with each other, and about the existence of special elements with special properties. There are eleven principal axioms, but you will observe, as we proceed, that the first ten occur in pairs, one for addition and a similar one for multiplication. In fact, note that, in most cases, the paired axioms are almost identical except for the notation of multiplication on the one hand and addition on the other.

CLOSURE AXIOMS

For every x and y, $z = x + y$ is a number.

For every x and y, $z = x \cdot y$ (written xy) is a number.

Observe that each of these is no more nor less than the statement that the sum or product of two numbers is again a number.

COMMUTATIVE AXIOMS

For every x and y, $x + y = y + x$.

For every x and y, $xy = yx$.

These merely state that the order of multiplying or adding two numbers is immaterial. We get the same answer either way.

ASSOCIATIVE AXIOMS

For every x, y, and z, $x + (y + z) = (x + y) + z$.

For every x, y, and z, $x(yz) = (xy)z$.

These axioms concern grouping. The point here is (as always in mathematics) that whenever parentheses (or any of their variations, such as braces or brackets) are used in this way, they mean that whatever is inside is to be considered as a single entity (here, a number).

The left side of the addition axiom says to add x to the result of adding y to z. The right side says to add z to the result of adding x to y. The axiom tells us that it doesn't make any difference how we group things; we always get the same result. Observe that the *order* of adding is not involved in this axiom.

This set of axioms also has an important practical application. Addition and multiplication are operations which are each defined only for a *pair* of numbers; that is, we can really add or multiply only two numbers at a time. This is the axiom which extends our ability to the point where we can deal with whole strings of numbers, not just the two which can be treated from the definition. It tells us that, no matter how we collect the elements of the string into bunches, we always get the same answer. Ordinarily, we add the first two numbers, and then add the third to the result, and the fourth to that result, and so on; but observe that we need not proceed only in this way.

UNIT AXIOMS   *There is a number (called zero and written 0) for which x + 0 = x for every number x.*   *There is a number (called one and written 1) for which x · 1 = x for every x.*

Zero and one are distinct.

Note the form of these axioms. They are called *existence axioms*, and they postulate that the special numbers zero and one, which we discussed in the last chapter, exist in algebra and retain their neutral qualities with respect to the symbols of algebra.

The phrase "zero and one are distinct" appears odd; yet, surprisingly, it is not possible, on the basis of the other axioms, to prove that one is not zero. In fact, the set which contains zero as its only element satisfies all the other nine axioms as well as the unit axioms with the phrase removed.

Also observe that nothing is said about there being only one neutral element for each operation. At least as far as the axioms are concerned, there might be several. That this is not the case is easily shown by the following theorem:

THEOREM 3.1   *There is only one neutral element for addition.*

Proof   Suppose there are two zeros (if there were more than one, there would have to be at least two). Write one as z and the other as 0.

Then $z = z + 0$ Since 0 is neutral under addition;

but $z + 0 = 0$ Since z is neutral under addition;

therefore $z = 0$ Things equal to the same thing are equal to each other.

What we have shown, then, is that there can't be two different neutral elements under addition.

INVERSE AXIOMS

For every x there is a y for which x + y = 0. Such a y is called an additive inverse of x.

For every x which is not zero, there is a y for which xy = 1. Such a y is called a multiplicative inverse of x.

There are two things to observe about these axioms. First, notice that they make up the only matched pair of axioms so far which is not symmetric (because of the exception of zero on the right) with respect to multiplication and addition. In each of the other pairs we could exchange the words *addition* and *multiplication* and not really change the meaning of the axioms once we had adjusted the wording.

Also, these too are existence axioms, and as with the previous existence axioms, the matter of uniqueness must be explored. Of course, we don't have uniqueness in the same sense that we had in the unit axioms. From our experience we know that the additive inverse for 6 will not be the same as the additive inverse for -18. But what we can show is this theorem: *For every x there is a unique y so that x + y = 0.* The claim is not that a single y will do for all x's, but that once we have been given a particular x, the y will be uniquely determined. The argument for the proof goes as follows: Suppose for some x there are two additive inverses; call them y and w. From the definition of inverse, $x + y = 0$ and $x + w = 0$, but then $y + x = 0$ (why?). Also

$y = y + 0$	The neutrality of zero
$y = y + (x + w)$	$x + w = 0$
$y = (y + x) + w$	Why?
$y = 0 + w$	$y + x = 0$
$y = w$	Neutrality of zero

This demonstrates that for any x there is only one additive inverse. Similarly, it is just as easy to show that for any $x \neq 0$ there is only a single multiplicative inverse. Observe that once we have proved these (and other) assertions from the axioms, they become part of the facts of algebra.

Once we have our uniqueness, we can introduce the familiar notations of $-x$ for the *additive* inverse of x and x^{-1} or $1/x$ for the *multiplicative* inverse of x. That is, $-x$ is that unique number which, when added to x, gives 0, while $1/x$ is that unique number which, when multiplied by x, gives one.

DISTRIBUTIVE AXIOM $$x(y + z) = xy + xz$$

In some ways this axiom is unique. It has no counterpart; that is, if we interchange the roles of addition and multiplication, we would get a statement which looks like $x + (yz) = (x + y)(x + z)$ which is, since it is to hold for all x, y and z, absurd. For example, it would state that $1 + 2 \cdot 3 = (1 + 2)(1 + 3)$, or that $7 = 12$. If we actually give the distributive axiom its full name, it tells us that *multiplication is distributive with respect to addition*. The absurdity of the example immediately above warns us that addition is *not* distributive with respect to multiplication.

However, the most important feature of this axiom is that it tells us how multiplication and addition interact with each other, and it is the only axiom that does this. As we have observed earlier, each of the previous axioms is concerned essentially with addition or multiplication.

We observe that we can easily show that $(x + y)z = xz + yz$, since $(x + y)z = z(x + y) = zx + zy = xz + yz$. This gives us both a *right distributive law*—just above—and a *left distributive law*—the axiom. We will refer to them jointly as the distributive law, without bothering to distinguish between the two.

As an illustration of the power of the distributive axiom, we will use it to prove that $(-x)y = -(xy)$. To this end, we consider $xy + (-x)y$. If we can show that this expression is zero, we will have shown that $(-x)y$ is *an* additive inverse of xy; but since there is only *one* such inverse, the two inverses must be equal; that is, we must have

$$(-x)y = -(xy)$$

Thus,

$xy + (-x)y = (x + (-x))y$	Distributive law
$xy + (-x)y = 0 \cdot y$	Inverse law, $x + (-x) = 0$
$xy + (-x)y = 0,$	0 times anything is zero. (See Exercise 3.2.13.)

so that

$$(-x)y = -(xy)$$

For convenience and ease of reference, we summarize (and condense) the axioms.

$x + y$ is a number	xy is a number	Closure
$x + y = y + x$	$xy = yx$	Commutative
$x + (y + z) = (x + y) + z$	$x(yz) = (xy)z$	Associative
$x + 0 = x$	$x \cdot 1 = x$	Unit
$x + (-x) = 0$	$x\left(\dfrac{1}{x}\right) = 1$	Inverse
	$x(y + z) = xy + xz$	Distributive

EXERCISES 3.2

1. Would it make any sense to formulate an axiom stating that subtraction is a commutative operation? Why? Since axioms cannot be proved, how do you justify your answer?

2. When we write down the associative axiom, we are already assuming one of the earlier axioms. Which one, and where do we use it? Could we do without it?

3. Restate the multiplicative associative axiom in words similar to those used in the text for the additive associative axiom.

4. Would the set of symbols $x + y + z$ have any meaning if the associative axiom were invalid? Explain.

5. What axiom (or axioms) are we using when we check the addition of a column of three or more figures by adding in the reverse direction?

6. Why isn't it necessary to include the statement "$0 + x = x$ for every x" in the units axiom?

7. Show that one is the only neutral element under multiplication.

8. Why is it not necessary to include in the inverse axioms "$y + x = 0$?"

9. Show that if $x \neq 0$, then there is only one y with the property that $xy = 1$. Would this still be true if the condition $x \neq 0$ were dropped? Explain.

10. We are all aware that $-x$ is numerically equal to $-1 \cdot x$ for all x, yet they are quite different in concept. What is the difference?

11. Explain why it would not be reasonable to use $-x$ for the additive inverse of x until we had shown uniqueness.

12. To show that $-(-x) = x$, we simply observe that $-x + x = 0$. Where in this have we used the uniqueness of the inverse?

13. Use the fact that $0 + 0 = 0$ to show that $0 \cdot x = 0$ for every x. (*Hint*: Multiply both sides of the expression by x; then use an inverse axiom.)

14. How do we know that addition is not distributive with respect to addition except for the fact that "it looks funny"?

15. Show that $(-x)(-y) = xy$. You can either follow the model in the text or use your imagination.

16. Can you now prove that $-x = -1 \cdot x$ for all x? (See Exercise 10.)

17. In high school you were often told to "multiply before you add unless there are parentheses." What axiom is this an expression of? Can you explain why?

3.3 Algebra Is a Language

We have all heard, to the point of banality, that mathematics is the language of science—and so it is. But what is the language of mathematics? What medium do mathematicians use to communicate their results or their ideas? Indeed, how are these ideas formulated? More often than not, algebra is the vehicle used for these purposes.

Like all languages, algebra has sentences that have their own form and structure, and that are used in different ways to convey different meanings. Probably the type of algebraic sentence you have met most frequently is the interrogative sentence, which you have called the equation.

If we consider a sentence of the form $3x - 12 = 9$, we see that we are really asking ourselves two questions: First, are there any numbers x which will make this a true statement, and second, if so what are they? Often we will answer the first part by actually finding a solution and displaying it, as in this case. Here when we are asked, "Are there any solutions?," we can say "Yes," since we know that 7 makes this true. But sometimes we run into problems which have no solutions. For example, it makes no sense at all to ask, "What are the solutions to $x^2 + 1 = 0$?" unless we have first asked (and answered) the question "Are there any solutions to $x^2 + 1 = 0$ at all?" If there are none, we can't very well ask, "What are they?" The answer in the number system we work in is "There are none." (To see this, note that x^2 is at least zero no matter what value x represents, while 1 is greater than zero, so their sum must be always strictly greater than zero.)

There are other kinds of interrogative sentences in algebra besides the equation; the inequality is but one example. Suppose we wish to find those values of x (if any) for which $\sqrt{x - 3}$ exists. Since the square-root operation is defined only for numbers which are nonnegative, we see that we are led to an inequality of the form $x - 3 \geq 0$. (Read this as "$x - 3$ is greater than or equal to zero.") Since we can add 3 to both sides of the inequality without changing its sense, we have $x \geq 3$. That is, for any x which is at least as large as three, we can find $\sqrt{x - 3}$.

Sometimes we ask several questions at once. For example, we can ask if there are any x's and y's which would make $2x + y = 3$ and $x + 2y = 1$ *both* true. In this case, we also show that the answer is "Yes" by displaying the correct x ($= 5/3$) and the correct y ($= 1/3$). Later on, we will show that, in this case, there is only one x and one y. However, if we ask "Are there any x's and y's which will make both $2x + y = 0$ and $2x + y = 1$ true?," we can see at a glance that the answer *must* be "No," since that would mean that one would be equal to zero, since things equal to the same thing are equal to each other. For future reference, we observe that nothing essential is changed if we multiply both sides of one or the other of these equations by the same constant.

Not all sentences of algebra are questions; some are declarative sentences. For example, there is no question involved in the expression $x^2 - y^2 = (x + y)(x - y)$. It is a statement, an assertion, which explains that for all x and y we can decompose (factor) $x^2 - y^2$ into the product of $(x + y)$ by $(x - y)$.

One of the most important uses of declarative statements is to tell us about one variable in terms of another. For example, if I throw a ball up into the air from ground level at 100 feet per second, then the expression $d = -16t^2 + 100t$ gives us the distance above ground (d) measured in feet, at any time (t) measured in seconds after it was thrown. Thus, after 3 seconds, the ball is $-16(9) + 100(3) = -144$

$+\ 300 = 156$ feet above ground. Observe that this is not an equation (as we are using the term), because it asks no questions; instead it imparts information. Often such a statement is called a *formula*.

Of course, once we have the device for generating information, we can ask questions of it. Perhaps we want to know when the ball hits the ground. This is the same thing as asking for those values of t which make d zero, since hitting the ground is the same as saying that the distance above the ground is zero. Setting $d = 0$, we *now* have turned our statement into the equation (question) $0 = -16t^2 + 100t$, which has as its solutions $t = 0$ and $t = 6\frac{1}{4}$. We can discard the solution $t = 0$ as being meaningless as a solution to our particular problem (why?), and have left the other solution, the ball will hit the ground $6\frac{1}{4}$ seconds after it is thrown upwards.

One of the principal uses of the declarative sentence in algebra is to express complicated ideas in a manner which is both concise and unambiguous. Let's try an algebraic expression without symbols first. What do we mean by *the sum of one variable and another squared*? We can't even begin to straighten this out unless we resort to algebra. There are, in fact, two perfectly good interpretations, either $x + y^2$ or $(x + y)^2$ and from the words there is no reason to prefer one above the other. Of course, we could reword the English so that the meaning becomes clear, but observe that in a properly constructed algebraic statement no ambiguity is possible.

The conciseness and clarity of algebra allows us to consider expressions of extreme complexity and even perform manipulations on them in a way which would never be possible if we were limited to words alone. Can you imagine writing out

$$\frac{\dfrac{\sqrt{x^2 + 2hx + h^2} + x + h}{\sqrt{x + h} - 1} - \dfrac{\sqrt{x^2} + x}{\sqrt{x} - 1}}{h}$$

in words? Yet it is a legitimate expression, and it is not unreasonable to suppose that a mathematician might wish to convert it into the equivalent statement

$$\frac{x^2 + hx - 2x - h - 1}{\sqrt{x + h} - 1\ \sqrt{x} - 1\,(\sqrt{x^2 + 2hx + h^2} + x + h\ \sqrt{x} - 1 + \ \sqrt{x^2} + x\ \sqrt{x + h} - 1)}$$

as someone who has been exposed to differential calculus might recognize. But it is inconceivable that the necessary manipulations could be carried out using only *words*, even supposing that the two expressions could be written in a reasonable way.

1. Determine which of the following have solutions. For those that do, see if you can determine what those solutions are. For those which do not, see if you can determine why not.

   (a) $\dfrac{1}{w-1} = 0$ (b) $t^2 + 4 = 0$

   (c) $\pi y = 3 - \sqrt{2}$ (d) $\dfrac{1}{z^2 + 1} = -1$

   (e) $\dfrac{x-2}{4x^2 - 3x - 2} = 1$ (f) $(z-3)^2 = -1$

2. Find those values of x, if any, for which the following square roots are defined.

   (a) $\sqrt{x-5}$ (b) $\sqrt{x+1}$ (c) $\sqrt{x^2 + 1}$ (d) $\sqrt{-x^2}$

3. What, if any, is the meaning of the solution $t = 0$ of the problem $0 = -16t^2 + 100t$ in the text?

4. Obtain the solution $6\frac{1}{4}$ for $0 = -16t^2 + 100t$.

5. Does the expression $d = -16t^2 + 100t$ of the text have any meaning for t greater than $6\frac{1}{4}$ or less than zero? Explain.

6. Which of the following are questions and which are statements? (*Note*: Since these are without context, some may be interpreted either way. Give both interpretations where there are two.)
   (a) $c = \pi d$ (b) $(x-2)^2 = x^2 - 4x + 4$
   (c) $x^2 \geq 0$ (d) $x^2 + y^2 - 1 = 0$
   (e) $v = xyz$ (f) $A = \frac{1}{2}bh$

7. Express the following in symbols. Where there is an ambiguity, express all possibilities.
   (a) The sum of two variables cubed.
   (b) The distance traveled upward by a thrown object is -16 times the time squared added to the initial velocity times the time.
   (c) The area of a trapezoid is the product of the length of the line segment joining the midpoints of the opposite nonparallel sides and the perpendicular distance between the parallel sides.
   (d) The square root of the square of a variable plus twice the variable added to one.

8. Write out the following in words without ambiguities insofar as possible, and try to keep them short.
   (a) $(1 + 2 + 3 + 4 + \cdots + 100)^2$
   (b) $1^2 + 2^2 + 3^2 + \cdots + 100^2$

(c) $\sqrt{x^2 + 2x + 1}$
(d) $\sqrt{x^2} + 2x + 1$
(e) The first displayed expression on page 61.
(f) The second displayed expression on page 61.

9. The instructions on federal income tax forms are difficult to read. Why? What does that have to do with Exercise 8?

3.4 Sense and Nonsense

When we are constructing words or sentences in English, we are not allowed to string together any letters or words in just any order we wish. There are certain rules we must follow if the resulting expression is to make sense. Every word, indeed every syllable, must have a vowel. Every sentence should have a verb, and so on.

We have similar structures to watch out for in algebra. Here the problem is somewhat complicated by the fact that some of the symbols we deal with can assume varying values, a difficulty not met with in English, where every letter has pretty much a fixed meaning. Nevertheless, we must always take care that our algebraic sentences make sense.

To begin with, we must make sure that we never divide by zero. As we have already seen, this one error could invalidate an entire sequence of operations. Thus, the expression $1/(x - 1)$ is meaningless if by some mischance we allowed x to assume the value one.

The set of allowable values that a variable (or the variables, if there are several of them) may assume for a particular expression is called the *natural domain* of that expression, so the natural domain of $1/(x - 1)$ is the set of all real numbers except 1. (Recall that all our symbols stand for nothing but real numbers.) Note that when we try to get any information out of $1/(x - 1)$, we get none when x is one.

When we look at an expression such as

$$y = \frac{x - 2}{x^2 - 4}$$

we can see that both 2 *and* -2 are not in its domain. It is tempting to note that $(x - 2)$ is a factor of both the numerator and the denominator, and hence to conclude that y is also equal to $1/(x + 2)$; but we hasten to point out that this process is allowed by our axioms only when x is *not* two. Thus the expressions

$$\frac{(x-2)}{(x^2-4)} \quad \text{and} \quad \frac{1}{(x+2)}$$

are not quite the same, since the $1/(x+2)$ includes the number 2 in its domain, while $(x-2)/(x^2-4)$ does not. It *is* true that, for those values of x for which they both have meaning, they are equal, but otherwise they are not.

It also can happen that we may get into trouble because we are asked to perform other impossible mathematical operations. Commonest of these is the taking of the square root of a negative number. This is not possible in our number system, since a negative times a negative is a positive; hence a negative number can never be written as the product of numbers which have the same sign. Thus, as we saw in the last section, the domain of the expression $\sqrt{x-3}$ is the set of all numbers which are greater than or equal to three.

We also observe that if we have an expression made up of several parts it is defined only when all of its individual parts are. Thus,

$$\frac{1}{x-4} + \sqrt{x-3}$$

has as its domain the set of all numbers which are greater than or equal to three *except* four.

When we are dealing with the application of mathematics to real problems, there are other precautions we must take. Consider the expression $d = -16t^2 + 100t$ which was the number of feet (d) above the ground an object would be t seconds after it had been thrown upward at 100 feet per second. We saw that there were two solutions to the equation $-16t^2 + 100t = 0$; that is, $d = 0$ for two values of t. This indicated that the object was on the ground twice, the first time at the instant it was thrown up, and the second time when it returned to earth $6\frac{1}{4}$ seconds later.

Now, if we look carefully at $-16t^2 + 100t$ purely as a mathematical expression without considering its background, there is no reason to suppose that it isn't defined for all values of t. But this is not just an abstract expression; it is also a *mathematical description of a physical event*, and *as such* it has no validity for t greater than $6\frac{1}{4}$ nor for t less than zero. That is, as a representation of a physical event, its domain is the set of all t between 0 and $6\frac{1}{4}$ inclusive, even though it may have a natural domain consisting of all real numbers t. Similarly, $A = \frac{1}{2}ab$ (as the formula for the area of a triangle) is meaningful only for positive a (altitude) and b (base), even though it has formal meaning for all values of a and b.

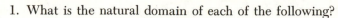

EXERCISES 3.4

1. What is the natural domain of each of the following?

(a) $\dfrac{1}{x-2}$ (b) $\dfrac{1}{x+6}$ (c) $\sqrt{x+1}$

(d) $\sqrt{-x^2}$ (e) $\dfrac{1}{\sqrt{x^2}}$ (f) $\dfrac{1}{\sqrt{-x^2}}$

(g) $\sqrt{x^3}$ (h) $\sqrt[3]{x^2}$

(i) $\dfrac{1}{x-1}+\dfrac{1}{x+1}$ (j) $\sqrt{x-1}\sqrt{x+1}$

(k) $\dfrac{1}{\sqrt{x-3}}$ (l) $\dfrac{x-1}{\sqrt{x-1}\sqrt{x+1}}$

2. Are the expressions $\sqrt{xy}$ and $\sqrt{x}\,\sqrt{y}$ equal? Why?

3. Consider the following algebraic expressions in the indicated contexts, and try to determine their domains in context. If there are several interpretations, try to give them all.
   (a) $c = \pi d$. Circumference of a circle is π times the diameter.
   (b) $c = 1.39p$. Cost of p pounds of hamburger at \$1.39 per pound.
   (c) $c = 100{,}000s + l + e$. The cost of a 10,000-square-feet-per-floor building is \$100,000 per story (s) plus the land cost (l) plus the cost of excavation (e).
   (d) $F = ma$. The directed force required (to change the motion of an object) is equal to its mass times the acceleration. (Assume all motion is along a straight line.)
   °(e) $QPA = \Sigma(G \times C)/T$. Quality point average is the sum (Σ) of grades received in each course (G)—expressed numerically—times the number of credit hours of that course (C), all divided by the total number of credit hours taken (T).

3.5 Algebra in Action

In this section we will solve three problems and obtain general solutions for each. These problems are not earthshaking. It is just that they indicate, in a very minor way, the type of approach taken by many modern mathematicians toward problem solving. You will probably know the results in each case, but the attitude may be new.

We consider first the most general linear equation in one variable, $ax + b = 0$. In this expression the letters a and b are fixed but undetermined or unspecified (with an exception, we shall note later). That is to say, we agree that although we don't know the values of a and b at this time, whatever values they do have will be unique and known to us for any particular problem. Thus the question $ax + b = 0$ is, Are there any values of x which will make this a true statement? Furthermore, if we can find a way to express x in terms of a and b, then we will have a solution, since we have supposed that for any particular problem, we will know a and b.

We first observe that, in spite of the generally unrestricted nature of a, we can at least assume that a is not zero. For, if a were zero, there would be no question left, and so, no equation, just the statement $b = 0$, which either is or is not true.

We can now proceed to attack $ax + b = 0$. Adding $-b$ to both sides gives $ax = -b$; and then (since a is not zero) we divide both sides by a and we will have $x = -b/a$. This, then, is our desired solution, since, if we replace x by $-b/a$, our original statement will be true. Thus we have shown, first, that if there is a solution it must be $-b/a$; and second, that $-b/a$ is a solution. These together tell us whenever a is not zero, the general linear equation must have a unique solution.

Notice that $-b/a$ is a general solution, in that it solves the linear equation once and for all. After we have put the equation in its standard form

$$ax + b = 0$$

we can read off a and b, and then, without going through any solution procedure, substitute the appropriate values into the expression $x = -b/a$. Furthermore, we now know, from our general investigation, that a solution always exists for this equation as long as a is not zero.

The next equation we will solve is the general quadratic, $ax^2 + bx + c = 0$. As before, the letters a, b, and c are fixed but undetermined, so any information we get about the solutions x will be in terms of a, b, and c.

To solve this equation, we first need to notice that any expression which can be written in the form $x^2 + rx + \frac{1}{4}r^2$ can be factored into $(x + \frac{1}{2}r)^2$, where here r is again fixed but undetermined. The use of this simple fact in solving for $ax^2 + bx + c = 0$ is one of the oldest techniques known in mathematics and is called "completing the square." We also observe that once again we may assume that a is not zero, for if it were we would have a linear equation, which we have already solved. Now we are ready to start.

$$ax^2 + bx + c = 0$$

Original equation

$$ax^2 + bx = -c$$
$$x^2 + (b/a)x = -c/a$$

Rules of equalities, inverses, and the fact that $a \neq 0$

$$x^2 + \frac{b}{a}x + \frac{1}{4}\frac{b^2}{a^2} = \frac{1}{4}\left(\frac{b^2}{a^2}\right) - \frac{c}{a}$$

Here is where we complete the square. Starting with $x^2 + (b/a)x$ we notice that if we add $\frac{1}{4}b^2/a^2$ to the left side we will have an expression of the form $x^2 + rx + \frac{1}{4}r^2$ with $r = b/a$. Since we added to the left we also must add the same thing to the right.

$$\left(x + \frac{1}{2}\frac{b}{a}\right)^2 = \frac{1}{4}\left(\frac{b^2}{a^2}\right) - \frac{c}{a}$$

We factor the left side. It was precisely to do this that we added $\frac{1}{4}(b^2/a^2)$ to the left side.

$$(x + b/2a)^2 = \frac{b^2 - 4ac}{4a^2}$$

We multiply and add the appropriate fractions.

or

$$\left.\begin{array}{c} x + \dfrac{b}{2a} = \sqrt{\dfrac{b^2 - 4ac}{4a^2}} \\[3em] x + \dfrac{b}{2a} = -\sqrt{\dfrac{b^2 - 4ac}{4a^2}} \end{array}\right\}$$

This is a purely formal operation. It cannot be meaningful if $b^2 - 4ac$ is negative. We will discuss $b^2 - 4ac$ and the problem of roots later on. Also note that any positive number has two square roots.

$$x + \frac{b}{2a} = \pm\sqrt{\frac{b^2 - 4ac}{4a^2}}$$

This is merely a new way of writing the last two bracketed statements on the left as one.

$$x + \frac{b}{2a} = \frac{\pm\sqrt{b^2 - 4ac}}{\sqrt{4a^2}}$$

Here we use the fact that the quotient of the square root is the square root of the quotient; see Exercise 2.4.11.

$$x + \frac{b}{2a} = \frac{\pm\sqrt{b^2 - 4ac}}{2a}$$

Since the denominator is a perfect square

$$x = \frac{-b}{2a} \pm \frac{\sqrt{b^2 - 4ac}}{2a}$$

Properties of equality, inverses, and units

$$x = \frac{-b \pm \sqrt{b^2 - 4ac}}{2a}$$

We can add the fractions, since the denominators are the same.

Here, then, are our "solutions," the *quadratic formula*, with all the information we need contained in it. If we replace x by either of the two values, we will get a true statement, at least formally. But it should also be clear that the key to this expression is under the radical, that is, $b^2 - 4ac$. This is so important it has a name, the *discriminant*. Now if the discriminant is negative, there are *no solutions* to our equation, since we cannot take the square root of a negative number; if the discriminant is zero, there is a *single solution*, namely $-b/2a$, since $\sqrt{0} = 0$; finally, if the discriminant is positive, we get *two distinct solutions*, one corresponding to the positive square root and one corresponding to the negative root.

Note that we can get the information about the number of solutions without actually solving our equation. For example, if we wish to know whether there are solutions for $3x^2 - 7x + 5 = 0$, we observe that here we have $a = 3$, $b = -7$, and $c = -1$, so that

$$b^2 - 4ac = (-7)^2 - 4 \times 3 \times 5 = 49 - 60 = -11$$

which tells us that there are no solutions. From this, we can then infer, for example, that the natural domain of $1/(3x^2 - 7x + 5)$ is the set of all real numbers, since the denominator can never be zero.

Finally, we turn to the general solution of two equations in two unknowns, which we write as

$$rx + sy = t$$
$$mx + ny = p$$

In this case the question is "Are there any *pairs* of values x and y which will make both of these statements true at the same time?" As before, we will assume that r, s, t, m, n, and p are all fixed but undetermined constants. Therefore, if we can express x and y in terms of the other letters, we will have solved our problem.

$$rx + sy = t,$$
$$mx + ny = p$$
Original equations

$$(nr)x + (ns)y = nt$$
$$(sm)x + (sn)y = sp$$
Multiply the top equation by n and the bottom by s, and use the associative law.

$$(nr)x - (sm)x = nt - sp$$
Subtract the bottom equation from the top.

$$(nr - sm)x = nt - sp$$
Factor.

$$x = \frac{nt - sp}{nr - sm}$$
Divide both sides by $(nr - sm)$, provided it is not zero.

Also

$$y = \frac{rp - mt}{nr - sm}$$
Steps similar to those above

Again, substitution of the values just found for x and y in the original equations will make both statements true.

From this we see that if the condition $nr - sm \neq 0$ is met, then there will be a unique pair of numbers x and y which will satisfy our set of equations.

But what if $nr - sm$ *is* zero? Then we have

$$nr - sm = 0$$
Original condition

$$nr = sm$$
Add sm to both sides.

$$\frac{r}{m} = \frac{s}{n}$$
Divide by n and m, if neither is zero.

Say $\frac{r}{m} = \frac{s}{n} = k$
Calling k the *common ratio*

$$\frac{r}{m} = k,$$
Rewriting

$$\frac{s}{n} = k$$

$$r = mk,$$
$$s = nk$$
Multiplying the equation on top by m, and on the bottom by n

$$(mk)x + (nk)y = t,$$
$$mx + ny = p$$

Substituting in the top of the original two equations for r and s

$$(mk)x + (nk)y = t,$$
$$(km)x + (kn)y = kp$$

Multiplying the second equation by k

$$0 = t - kp$$

Subtracting the lower equation from the upper

What we have shown thus far is that the condition $nr - sm = 0$ (or equivalently, $nr = sm$) means that $t - kp = 0$ whenever there are any pairs x and y which make both equations true statements. That is, in order to obtain any solutions to our original equations, we must have $t = kp$ whenever $r/m = s/n = k$ (notice that the k's are the same in all three instances); otherwise there are no solutions.

But what happens if both $nr - sm = 0$ and $t = kp$ are true? In this case (substituting kp for t in the next-to-last step), we have

$$(mk)x + (nk)y = kp$$
$$(mk)x + (nk)y = kp$$

That is, we have only a single equation. Solving, say, for x, we obtain

$$(mk)x = kp - (nk)y$$

$$x = \frac{kp - (nk)y}{mk}$$

or

$$x = \frac{p - ny}{m}$$

We see from this that there are infinitely many solutions. All we have to do is pick *any* value for y, put it into the last formula, and x is then determined. Substitution shows that any pair x and y so constructed is a solution.

We still need to clean up one more detail. Suppose $nr = sm$ and n is zero; then $nr = 0$, and so also $sm = 0$, which, in turn, means that either $s = 0$ or $m = 0$. But if both n and s are zero, the unknown y disappears from the pair of equations, while if n and m are both zero, there is *no second equation*. We have treated each of these cases earlier, so we can ignore the possibility that $n = 0$. Similarly, we need not worry if $m = 0$.

To summarize: If we start with the pair of equations

$$rx + sy = t$$
$$mx + ny = p$$

then there will be a unique pair of numbers x and y which will make both true at the same time whenever $rn - ms$ is not zero. Whenever $rn - ms = 0$, then there is *no unique* solution; either there are no solutions at all, or there are infinitely many pairs x and y which make both equations true.

These discussions serve to illustrate in a very small way one of the triumphs of modern mathematics. There is an important branch of mathematics, differential equation theory, which is of great practical importance to physicists and engineers. Most equations of this theory cannot be solved; yet for many of these, mathematicians have been able to discover significant facts about the solutions even though they don't know what these solutions are.

EXERCISES 3.5

1. Use the general solution to write down, without explicitly solving, the solutions to $ax + b = 0$.
   (a) $3x + 5 = 0$ (b) $2x = -5$

   (c) $3x = 7$ (d) $1 + x = \dfrac{5}{3}$

   (e) $\pi - \pi x = \sqrt{2} - \dfrac{3}{\pi}$

°2. Explain how we could also have obtained a general solution to $ax + b = 0$ by assuming that $a = 1$.

3. Determine which of the following have solutions by examining the discriminant.
   (a) $3x^2 + 7x + 5 = 0$ (b) $x^2 + 7x + 5 = 0$
   (c) $3x^2 - 7x - 1 = 0$ (d) $7x^2 = 3x + 1$
   (e) $6x + 2 = (9/2)x^2$ (f) $1 - 2x^2 = x$
   (g) $\sqrt{2}x - x^2 = -12$

4. Use the general solutions we obtained in the text to find the solutions to those problems in Exercise 3 which have them.

5. Is it possible to have exactly one solution to $x^2 + bx + c = 0$ if b is an odd integer and c is any integer? Explain.

6. Carry out the steps to obtain the solution for y on page 69 in the text.

7. Determine which of the following pairs of equations have unique solutions, nonunique solutions, or no solutions.

(a) $4x + 3y = 2$
    $3x + 4y = 1$

(b) $2x - y = 1$
    $-x + \frac{1}{2}y = 3$

(c) $-2x + y = 1$
    $4x - 2y = -2$

(d) $x - y = 1$
    $x + y = 1$

REFERENCES

Cajori, Florian. *A History of Elementary Mathematics*. Macmillan Publishing Co., Inc., New York, 1924.

Cajori, Florian. *A History of Mathematical Notations*. 2 vols. Open Court Publishing Company, La Salle, Ill., 1928.

Cardano, Gerolamo. *The Great Art, or the Rules of Algebra*. Translated and edited by T. Richard Witmer. The M.I.T. Press, Cambridge, 1968.

Hogben, Lancelot. *Mathematics for the Millions*, 4th ed. W. W. Norton & Company, Inc., New York, 1968.

Hooper, Alfred. *Makers of Mathematics*. Random House, Inc., New York, 1948.

Kline, Morris. *Mathematics and the Physical World*. Doubleday Publishing Company, Anchor Books, Garden City, New York, 1963.

Newman, James R. *The World of Mathematics*, 4 vols. Simon & Schuster, Inc., New York, 1956.

Whitehead, Alfred North. *An Introduction to Mathematics*. Oxford University Press, New York, 1958.

Chapter 4

Geometry

4.1 Introduction

Of all mathematics, as we understand and are using the term, the oldest is geometry. The first recorded geometer, a Greek called Thales, was active about 600 B.C. Since his time the interest in geometry has waxed and waned, now flourishing at the forefront of research mathematics, now hibernating in relative obscurity. Yet in all this time geometry—more properly plane Euclidean geometry—has defied all attempts to replace or supersede it.

For centuries, in fact, geometry was considered to be the very epitome of truth, to the point where the German philosopher Immanuel Kant (1724–1804) felt that it was the only yardstick of truth.

That is, the truth or falsity of any statement could only be judged against a background of Euclidean geometry; the degree to which a statement could be made to fit into a geometric context determined the level of confidence in the judgment of the truth value.

Of course, today we no longer hold this view. As we have already seen, the truth of a mathematical system depends upon the truth of its axioms, and these later can only be verified inductively. Thus we can never know if Euclidean geometry is "true," only that it seems to give us a very good way of describing certain situations in nature.

Perhaps this is the secret of plane geometry's incredible longevity. There is no question that many of the results of geometry have an enormous intuitive appeal. There is a considerable satisfaction in putting together a complex and sophisticated abstract structure, and then in having many of its elements correspond so closely to the natural observations we make.

There is yet another reason for our not accepting Euclidean geometry as the only form of truth. There exist other geometries! That is, there are other geometrical systems based on axioms which differ from those of Euclid, usually in only one or two specifics. Furthermore, these geometries have some properties which may be intuitively repugnant to some of us. In one of these, it can be shown that there are *no* pairs of parallel lines anywhere in the system, while in another it can be shown that given a line L and a point p not on L, there are actually *infinitely many* lines through p parallel to L. Yet these geometries also can be used to accurately describe situations which occur in nature, even if their corresponding realities are not as common as those of plane geometry. This can be explained, in a somewhat simplified way, by noting that Euclid's geometry is the geometry of the plane (or of flat surfaces). The others are geometries of surfaces which are curved, such as the surface of a sphere.

4.2 Euclid's Elements

Of all the books ever written on a nonreligious subject, undoubtedly the most widely used and most thoroughly studied is Euclid's *Elements*. Written about 300 B.C., it has remained the basic textbook of geometry ever since. Even today most high school geometry textbooks are fashioned after Euclid, using what is essentially his arrangement and order of presentation. True, corrections and improvements have been made, but what other book by a single author has continued in general circulation for over 2200 years? Clearly a work of genius.

We first observe that the *Elements* is just that; it was written as a textbook. It was not the first textbook on geometry; other earlier ones are known to have existed. It's only that the *Elements* was so superior to any of its predecessors that most of the latter just disappeared and their contents are lost to us. (Too bad. It would have been interesting historically to know what was being taught *prior* to Euclid.)

Euclid, furthermore, never intended the *Elements* to be a compendium of all known geometry. There are theorems which were in existence in Euclid's time which do not appear in the *Elements*. Euclid himself is supposed to have written a book on the conic sections (the parabola, the ellipse, and the hyperbola) but does not mention them in the *Elements*.

Euclid[1] used a total of ten axioms divided into five postulates and five common notions. The idea was that the postulates were peculiar to geometry, while the common notions were meant to hold for all of mathematics. Today we would make no such distinction.

The postulates, somewhat rephrased, are as follows:

1. A straight line may be drawn to include any two points.
2. A finite line segment can be produced in *either* direction to a straight line.
3. A circle can be drawn with any radius and with any point as center.
4. All right angles are equal to each other.
5. If two lines (L_1 and L_2 in Figure 4.1) are such that a third line (m) intersects them in such a way that the sum of the two interior angles (a and b) on one side of M is less than two right angles, then the two lines, if produced sufficiently far, will meet on the same side of the third which has the sum of the interior angles less than two right angles.

Figure 4.1

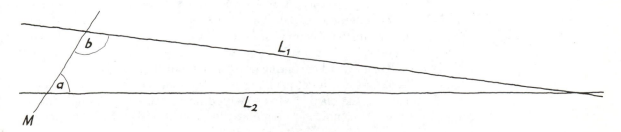

[1]This discussion of Euclid's *Elements* is based primarily upon Thomas L. Heath, *The Thirteen Books of Euclid's Elements*, Vols. I, II, III. Dover Publications, Inc., New York, 1956. Reprinted from Cambridge University Press edition, 1925.

The five common notions are:

1. Things equal to the same thing are equal to each other.
2. If equals are added to equals, the sums are equal.
3. If equals are subtracted from equals, the remainders are equal.
4. Figures which coincide with each other are equal.
5. The whole is greater than any of its parts.

There is more to these axioms than meets the eye, as we shall see. Nevertheless, the construction of such a magnificent edifice on the basis of so small a number of axioms is indeed an accomplishment of the highest order. It is true that the axioms are not complete; that is, Euclid did lean more upon his intuition and his diagrams than he realized (a matter we will discuss further on). Yet even with the additions which have been found necessary, the fundamental structure and content of the *Elements* has remained unchanged.

The first of the postulates insures, not only that there are such things as straight lines, but also that there are all the straight lines that will ever be needed. Further, there appears to be an implication of uniqueness; that is, given two points, there is only *one* line through them. Certainly Euclid tacitly assumed that such was the case.

The second postulate implies (although it does not state explicitly) that two distinct lines cannot share a common line segment; that is, the straight line so produced is understood to be unique. Note, also, that it seems to imply that the line segment can be extended as far as we wish, with no limit on its length. Again Euclid uses this assumption in his proofs.

The third is again an existence axiom. It gives us all sizes of circles in all locations. In addition it tells us something about the underlying plane. Since there is no limitation on the size of the radii, there can be no limitation on the size of the plane in any direction. Note, too, that there is no minimum radius either, which gives us a kind of continuity to the plane; that is, the plane cannot be a polka-dot-like discrete collection of points set at some minimum distance from each other.

In addition, postulate four is not as simple as it looks. It tells us that, no matter where we set a right angle down on the plane, it is the same size. Thus the right angle serves as some kind of invariant yardstick, a universal measuring device. Without the postulate there might be distortion of right angles as they were moved about from place to place on the plane, just as a twisted mirror distorts the objects it reflects. Also, note that without this postulate, the next one becomes meaningless.

Postulate five is a work of genius, and we defer the discussion of it until later. Of the common notions, there is no question that all—except perhaps number four—are indeed common. Observe that we are not asserting that they *are* true or that they can be proved true—just that they certainly are in accord with our experience. It also happens that there are mathematical structures in which number five would read that the whole is *no smaller* than any of its parts.

Common notion four is a little more complex than the others. In part it is a definition of equality (or congruence). Also, Euclid used it to justify the principle of superposition, the principle which allowed him to "move" figures about on the plane without change of shape or size. True, he did not use it often (he may have thought it inelegant), but it is fundamental to the whole of *Elements* since it is used to prove proposition 4 of book I, the famous side-angle-side theorem of triangle congruence. Thus, it probably belongs with the propositions as being geometrical in nature.

EXERCISES 4.2

1. To the Greeks, an axiom was a "self-evident" truth. What is self-evident about postulate 5?

2. Would it be desirable to include in the postulates, "6. Two straight lines cannot enclose space"? Why? (or why not?)

3. On the basis of the axioms alone, can you find out when a point is between two points? (*Hint*: This question is nowhere as innocent as it appears.)

4. The basic objects of Euclid's elements (straight lines, circles) stem from his visual experience. Suppose a race of intelligent fish developed a geometry.
   (a) What might the basic objects of such a geometry be?
   (b) Can you imagine some axioms of a watery geometry?
   (c) Would motion, perhaps, play a large part in such a geometry? Explain and contrast with Euclid's *Elements*.

4.3 The Fifth Postulate: Flatland[1]

Even to the most casual reader, the fifth postulate of Euclid's *Elements* appears to be quite different from the first four. The others are simple sentences, easy to state, and at least on the surface, easy to grasp. The

---

[1]With a low bow to Edwin A. Abbott for his delightful little *Flatland: A Romance of Many Dimensions*, Barnes & Noble Books, New York, 1963; a story of life in two dimensions.

fifth, on the other hand, is complicated to state, and certainly would be difficult to grasp without a diagram.

This complexity was deeply disturbing not only to the Greeks but to later mathematicians as well. There was a strong feeling that it was not an axiom at all but a theorem, which could be proved as a consequence of the other nine axioms. For over 2000 years it was the "scandal of mathematics" that no one was able to produce such a proof. There were many attempts, even by mathematicians of the highest caliber, yet no valid proof was ever constructed using *only* the other nine axioms. These efforts came to a halt in the middle of the nineteenth century, when three mathematicians working independently, showed that such a proof is impossible. Yet such was the quality of Euclid's genius that he saw this 2000 years before.

The false "proofs" of the fifth postulate which were constructed were not, as a rule, invalid in the usual sense. Instead, they failed because each contained at least one assumption which was not among Euclid's other nine axioms but which was entirely equivalent to Euclid's fifth.

This notion of equivalency is worthy of further exploration. We will say that two sets of axioms are equivalent if essentially the same collection of theorems can be proved from one set as from the other. In our case, we will hold all but the fifth axiom fixed, and substitute in various others for the fifth. Now suppose P is some theorem provable from Euclid's usual ten axioms, and also suppose that, after we substitute P for the fifth, the statement of the fifth axiom can be proved now as a theorem from the nine axioms and P. Then the two systems are said to be equivalent. That is, every proof which needed the fifth now uses the assumption P instead. In this case, we will also call P equivalent to the fifth.

There are, it turns out, many statements which are equivalent to Euclid's fifth. This is because this is the axiom which gives expression to the property of the plane which we call "flat," and there are many ways to describe this.

The equivalent statement most often used is called Playfair's axiom: Let L be any line and p be a point not on L; then there is *only* one line through p which is parallel to L (see Figure 4.2).

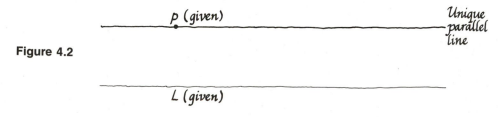

Figure 4.2

CHAPTER 4 / GEOMETRY

This is probably the postulate you learned in high school, and although easier to grasp than Euclid's form, at first, it is inferior to Euclid's as an axiom. For how do you even begin to verify it? Essentially, it tells you that something does *not* happen. Remember, two lines in a plane are parallel if they do *not* meet. Now, since lines can be produced indefinitely how can you ever tell for a certainty that the two particular lines you are interested in do not meet at some point which is beyond wherever you have decided to stop checking? Thus, it is impossible to determine by experiment whether just two lines are parallel. How much more difficult would it be to determine that there was *only one* line parallel to L?

Euclid's form, on the other hand, can be inductively verified more easily. All one has to do is measure the angles a and b and if their sum is less than $180°$, check to see whether L_1 meets L_2 on the side of m. If the sum is not too close to $180°$, there will be no difficulty in this. (See Figure 4.3.) Problems arise only as the sum nears $180°$.

Another equivalent statement is that the angle sum of every triangle (that is the sum of their three interior angles) is $180°$, or, even more astonishingly, that there is even one triangle whose angle sum is $180°$. Observe, then, that this asserts that if even one triangle has this property, they all do.

We list a few other alternative equivalent forms:

There exist similar triangles which are not congruent; that is, there exist two triangles, with the angles of one equal respectively to the angles of the other, which are not congruent.

The square on the hypotenuse of a right triangle equals the sum of the squares on the other two sides (Pythagorean theorem).

If a line intersects one of two parallel lines it *must* intersect the other.

Two lines parallel to the same line are parallel to each other.

Given any three points which are not collinear, it is possible to pass a circle through all of them.

If K is any counting number, there is a triangle whose area is larger than K.

Figure 4.3

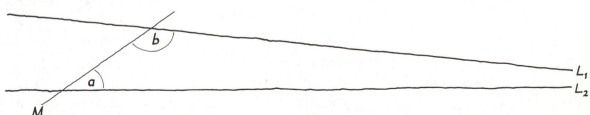

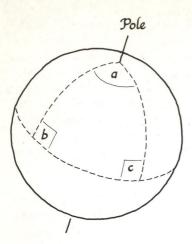

Figure 4.4

We repeat that any one of the above can be taken to be an axiom which, if it is inserted into Euclid's ten axioms in place of his fifth, will yield the same collections of theorems. Since the fifth characterizes the flatness of the plane, so must each of the others; this means they describe the real world only where it can be idealized by flatness.

As an illustration, consider a geometry on the *surface* of a sphere such as an idealized Earth. We take as our axioms the usual nine of Euclid, with the second postulate somewhat modified to allow for the fact that a line cannot have infinite length. "Lines" are taken to be arcs of certain circles.

Suppose, now, we consider a spherical triangle formed by the equator and two lines passing through the poles (longitudes). (See Figure 4.4.) Angles b and c will always each be right angles (90°), while angle a can be anything we like up to, but *not* including, 360°. Thus, on a sphere, we have produced triangles whose angle sums are anywhere from 180° to 540° (exclusive, in both cases).

We now know that there can be no triangles with an angle sum of exactly 180°, because if there were, then all triangles would have the same 180° angle sum. We can, in fact, show that every triangle on the sphere has an angle sum of more than 180°. This fact expresses the convexity of the sphere. We also know, on the basis of equivalence of axioms, that none of the statements about parallel lines will hold on this surface either; in fact there are *no* pairs of parallel lines. Circles which do not intersect are not among the "lines" of this geometry.

On the other hand, suppose we lived on a concave surface, for example on the outside of the bell end of a bugle. Mathematicians call an idealization of this a pseudosphere. Here we can take the nine axioms of Euclid without change, but what happens to the fifth? Again, it is simplest to look at the angle sums, and on the pseudosphere, a typical triangle looks like that in Figure 4.5.

It should not be hard to convince yourself that in this case the angle sum is *less* than 180°; and, in fact, all triangles on the pseudosphere share this property. (Recall that no angle sum can equal 180°.) So on the pseudosphere, as on the sphere, the fifth axiom cannot hold. The analogue of Playfair's axiom comes out like this on the pseudosphere: Let L be a line and p be a point not on L; then there are *at least two lines* through p parallel to L. (See Figure 4.6.) Of course they

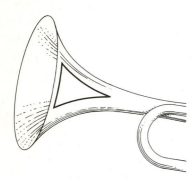

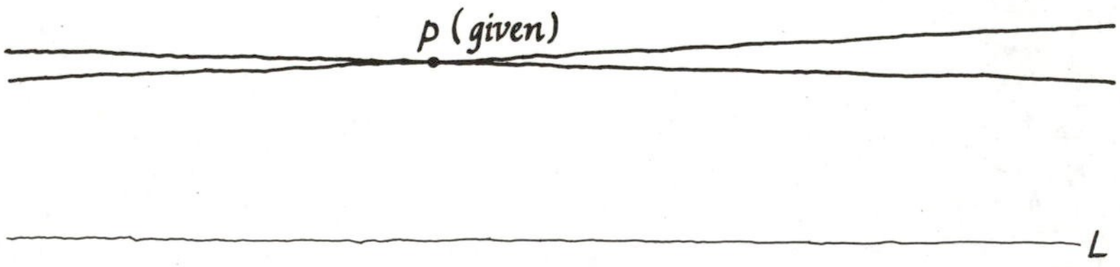

Figure 4.5

Figure 4.6

don't look parallel, but that is because the page of the book is like a flat plane, where we have only a single parallel line through p.

EXERCISES 4.3

1. Using diagrams like those in Chapter 1, show how Euclid's axioms, and the nine axioms plus the fifth replaced by a P, as described in the text, lead to the same valid deductive conclusions.

2. Go to a dictionary and look up the word *flat*. Check out the essential circularity in the definition. Is this a good undefined notion? Can you improve on it by expressing it in other terms?

3. Can you devise a scheme for verifying inductively when two lines are parallel?

4. Perhaps you can devise a method for verifying Playfair's axiom inductively. Check your method to see whether or not you are adding in assumptions.

5. In connection with Playfair's axiom, imagine that we live inside a spherical universe (that is, that our universe is like a gigantic orange).
   (a) What does a flat plane look like in such a universe? (*Hint*: What happens when you slice an orange?)
   (b) Would Playfair's axiom hold there?

6. Euclid proved that lines parallel in the plane do not meet. Comment.

7. Yet another alternative to the fifth is the statement "The line segment joining the midpoints of two sides of a triangle is one-half the third side." Prove that this implies one of the alternatives listed in the text.

4.4 Some Flaws in the Jewel

In view of the ancient age of Euclid's *Elements* and the vast numbers who have studied it, it would indeed be surprising if it had never been subjected to criticism. Of all the criticisms leveled at the *Elements*, the most justified is that its axioms are incomplete; that is, Euclid leaned too heavily upon his pictures, and for this reason, some axioms

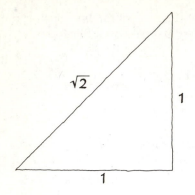

Figure 4.7

which should have been postulated were just assumed without explicit statement.

Before we go on, let us look at the consequences of relying too heavily upon pictures. We will pick an example (not from Euclid!) which "proves" by use of pictures that $2 = \sqrt{2}$! Consider an isosceles right triangle with legs equal to one. Then, by the Pythagorean theorem, the hypotenuse is equal to $\sqrt{2}$. (See Figure 4.7.)

On the other hand, consider the following constructions. We first bisect each side, and construct the first step-curve as in the illustration below:

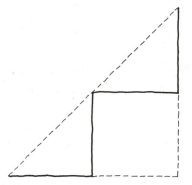

The length of all its vertical parts is 1, and the length of its horizontal parts is also 1; so its total length is 2. We now bisect each part of the first step-curve to get a second step-curve:

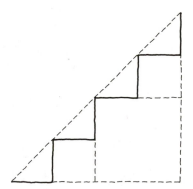

Again the vertical parts have length 1, as do the horizontal parts, so the second step-curve has length 2. Similarly we construct the third step-curve:

CHAPTER 4 / GEOMETRY

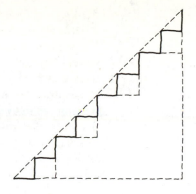

whose length is 2. We keep repeating the process and we see two things happening: The total lengths of each of the step-curves remains fixed at 2, while the curves themselves always keep coming closer to the hypotenuse. By continuing the process long enough we will get a crinkly step-curve the length of which is 2 and which is practically indistinguishable from the hypotenuse. Hence its length must be practically indistinguishable from the length of the hypotenuse, and if we have continued on long enough, it will equal the length of the hypotenuse. Thus $2 = \sqrt{2}$!

Of course, Euclid constructed no such paradoxes. We do not suggest, by our illustration, that he would have accepted such a contradiction; but the danger of accepting the evidence of our eyes alone is clear. Besides, we must again emphasize that what we (and Euclid) are proving are theorems not about the streaks and marks on a piece of paper, but about *abstractions* which are merely represented by the pictures.

The difficulty can be seen in the very first theorem of Book One, where Euclid constructs an equilateral triangle on a given line segment AB. He draws the two circles with centers at A and B and both with radius AB (see Figure 4.8), and then joins A and B each to the point C of intersection of the circles.

Figure 4.8

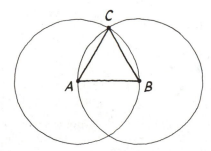

Simple enough! Foolproof? Not quite!

How do we know the point C actually exists? From the picture? Where in the axioms is justification for the conclusion that there are "enough" points so that, when we have the situation above, it won't be possible for one circle to "pass through" the other without actually intersecting? What is needed here is an axiom of "continuity," that is, an axiom which tells us that the points on lines and circles are dense enough so that there are *no gaps* between them.

Keep in mind that the truth of the conclusion is not, by itself, an indication of the validity of the argument. We must always be sure that our theorems follow as a deductive consequence of our axioms alone.

Another deficiency stems from the fact that there is no way of telling when one point is between two other points even when they are collinear, nor of telling what happens to a line which enters the vertex of an angle. Does the line cross one of the lines of the angle? What is needed here is an axiom of the type "if a line passes through a vertex formed by two sides of a triangle and through a point in the interior of the triangle, it intersects the third side." Of course we will have to define what we mean by the "interior of a triangle."

There have been attempts in relatively recent times to complete Euclid's axioms. Of these, the most celebrated is the set promulgated by the German mathematician David Hilbert (1862–1943) at the turn of this century. These are quite extensive, comprising sixteen axioms divided into five groups.

We have said nothing about Euclid's definitions, for they are generally poor and unworkable. He doesn't even seem to have made use of them himself. Moreover, he doesn't appear to have been aware of need for undefined notions, amazing as this seems. Modern treatments remedy this situation, starting with the point, plane, and line as being undefined.

EXERCISES 4.4

1. Make a large-scale drawing of the illustration of the step-curves and construct more, say, up to the sixth. Convince yourself that the assertions in the text about them are all correct.

2. Suppose we have two concentric circles and a segment containing a radius of each. Then we roll the two circles together along a

Figure 4.9

tangent to the larger one, until we have completed exactly one revolution, as in Figure 4.9. What does this picture "prove" about the circumferences of the two circles?

3. If you have done some graphing of curves in high school, consider the graph of the parabola $y = x^2$ and the graph of the line

$$y = 2x + 1$$

Where do these intersect each other in a mathematical structure with no irrational numbers? (*Hint*: "Things equal to the same thing, etc.," yields a quadratic equation to solve.)

4. Invent an axiom (or definition) which tells us when one point is between two others.

5. Invent a definition for the interior of a triangle. Also, do the same for the interior of an angle. What difficulties do you encounter in the latter definition not present in the first?

6. Suppose we are considering the geometry of the *circumference* of a circle. How could we define "a point is between two points"? Is the definition as simple as in Exercise 4? Why?

7. Euclid never made this mistake, but it has appeared, incorrectly, in many high school textbooks. It purports to be a proof that the base angles of an isosceles triangle are equal *using only the ten axioms of Euclid.* Can you find the error?

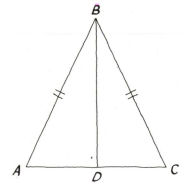

Figure 4.10

Given $\triangle ABC$ with $AB = BC$. (See Figure 4.10.) To prove $\angle A = \angle C$, construct BD to be the bisector of $\angle B$. (Such a construction is easily proved possible.)

Then $\triangle ABD$ is congruent to $\triangle BDC$ (S.A.S). Therefore, $\angle A = \angle C$. (Corresponding parts of congruent triangles.)

8. Try to give a proof of theorem in Exercise 7, with no constructions at all.

4.5 An Application: Trigonometry

Trigonometry is one of the earliest applications of Euclidean geometry. It was invented in order to deal with problems in astronomy, but wide uses have been found here on earth as well.

The basis of trigonometry is the theory of similar triangles. This means that, in a system in which Euclid's fifth postulate fails to hold, there can be no trigonometry (as we understand the term) since there are no noncongruent similar triangles.

We start by considering two right triangles (Figure 4.11) with an acute angle of one, A, equal, to an acute angle A', of the other; since $\angle C = \angle C'$ (all right angles are equal), we must have $\angle B = \angle B'$. Thus $\triangle ABC$ is similar to $\triangle A'B'C'$.

Note that we have shown that *any right triangle with one acute angle equal to A must be similar to ABC*. From the theory of similar triangles[1] we then know that

$$\frac{BC}{AB} = \frac{B'C'}{A'B'} \quad \text{and} \quad \frac{AC}{AB} = \frac{A'C'}{A'B'}$$

for every triangle $A'B'C'$ which is similar to triangle ABC. Therefore, and this is the point, the *ratios of opposite to hypotenuse and adjacent to hypotenuse depend only on the angle A* and not upon the particular right triangle we happen to be looking at. This is why we can call the ratio (*opposite*)/(*hypotenuse*) the sine of A (abbreviated *sin A*), and the ratio (*adjacent*)/(*hypotenuse*) the cosine of A (abbreviated *cos A*).

Let us observe how the number sin A behaves as the angle A itself changes. The easiest way to do this is to limit our inquiry to those triangles that have hypotenuses of length 1. This is no restriction, since we know that the sine of A does not depend on the particular triangles

Figure 4.11

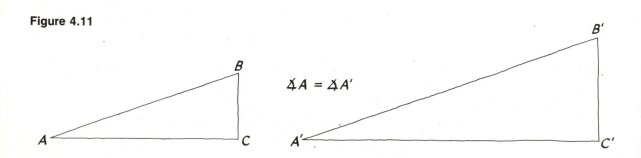

$\angle A = \angle A'$

---

[1]Recall that in similar triangles, corresponding pairs of sides are proportional.

CHAPTER 4 / GEOMETRY

Figure 4.12

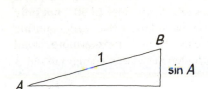
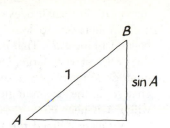
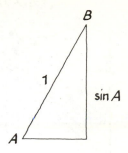

we have under consideration. When we look at these special triangles we see that the ratio

$$\frac{\text{(opposite)}}{\text{(hypotenuse)}} = \frac{\text{opposite}}{1} = \text{opposite}$$

That is, sin A is equal to the length of the opposite side.

Suppose we draw a few such triangles for different values of the angle A. Examination of the diagrams in Figure 4.12 leads us to conclude that, for small values of A, sin A is also small, while when A is larger, sin A is also larger. If we keep the point A fixed, as well as the placement of the side adjacent to A, then the point B will sweep out a quarter circle of radius 1 as we vary the angle A. (Recall that we consider only triangles with hypotenuse 1.) We will then get a picture which looks like Figure 4.13.

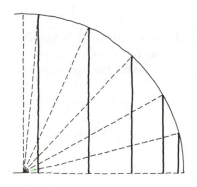

Figure 4.13

The dashed lines represent the different triangles we get for different values of A, while the length of the solid line is sin A. Thus we can actually trace what happens. As angle A grows from 0° to 90°, exclusive in both cases, sin A grows smoothly from zero to one, also exclusive in both cases. It is true we are leaning heavily on our picture, but these results can all be verified by other means (see Exercise 4.5.6).

Another interesting number associated with the angle A is the tangent of A (abbreviated *tan A*). It is no accident that this bears the same name as a line which intersects a circle only once, even though it is usually defined to be the ratio *(opposite)/(adjacent)*. To see the derivation of the name, we once again consider all possible triangles with hypotenuse one, and place them as in Figure 4.13. We also construct the tangent (line) T to the circle at E (see Figure 4.14). Then the triangles ABC and ADE are similar, so we don't need triangle ABC any longer, and

$$\tan A = \frac{\text{opposite}}{\text{adjacent}} = \frac{DE}{1} = DE$$

That is, the *tangent* of angle A is the length of the line segment cut off the appropriate *tangent* line by an extension of the hypotenuse of $\triangle ABC$.

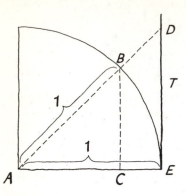

Figure 4.14

The analysis of the tangent now becomes a simple matter of looking at DE. For small sizes of A, the length of DE is also small; then as angle A increases, so does DE until, as angle A nears 90°, DE increases beyond all measure. That is, for values of A very close to 90°, not only does DE get very large, but it can be made larger than *any* counting number K by taking A sufficiently close to 90°. Furthermore, small changes in the size of angle A can produce very large changes in tan A when A is near 90°.

We have defined the trigonometric functions only for angles between 0° and 90°. It is possible to define them for other angles, as we shall see later on.

EXERCISES 4.5

1. The calculation of the actual values of sin A and cos A is usually a very difficult and tedious process and does not involve geometry at all. Nevertheless for certain special angles the calculations are easy.
   - (a) Use the properties of isosceles triangles to find sin 45° and cos 45°.
   - (b) Use the properties of equilateral triangles to find sin 60°, cos 60°, sin 30°, cos 30°.
   - (c) Use the results of b to determine whether
   $$\cos (A + B) = \cos A + \cos B.$$

2. The cotangent of A is defined to be (*adjacent*)/(*opposite*). Is it independent of the triangle selected? Why?

3. Is sin 0° defined? Could it be defined in a way which makes sense? If so, what would its value be? What about sin 90°? Does it make sense to ask what is sin 95°? Explain.

4. Show that the sine of A can never be greater than 1, by use of the Pythagorean theorem.

5. Analyze the cosine of A as A grows from 0° to 90°, using a figure like Figure 4.13.

6. For those who have had a little analytic geometry: Can you analyze sin A and cos A as we have done, using algebra? (*Hint*: Place A at the origin and use the equation of the unit circle, $x^2 + y^2 = 1$; then express sin A and cos A in terms of x and y.

7. Is there a reasonable way of defining tan 0°? What about tan 90°? Explain.

REFERENCES

Aaboe, Asgar. *Episodes from the Early History of Mathematics*. Random House, Inc., New York, 1964.

Beckmann, Peter. *A History of π*, 2d ed. Golem Press, Boulder, Colo., 1971.

Brydegaard, M., and J. E. Inskeep, eds. *Readings for Geometry from the Arithmetic Teacher*. National Council of Teachers of Mathematics, Washington, D.C., 1970.

Cajori, Florian. *A History of Mathematical Notations*. 2 vols. Open Court Publishing Company, La Salle, Ill., 1928.

Cundy, H. Martyn, and A. P. Rollett. *Mathematical Models*, 2d. ed. Oxford University Press, London, 1961.

Dantzig, Tobias. *The Bequest of the Greeks*. Charles Scribner's Sons, New York, 1955.

Gow, James. *A Short History of Greek Mathematics*. 1884. Revised reprint. Chelsea Publishing Co., Inc., New York, 1968.

Heath, Thomas. *A History of Greek Mathematics*. Oxford University Press, Inc., New York, 1960, Vols. I, II.

Heath, Thomas. *A Manual of Greek Mathematics*. 1931. Reprint. Dover Publications, Inc., New York, 1963.

Heath, Thomas L. *The Thirteen Books of Euclid's Elements*. Dover Publications, Inc., New York, 1956, Vols. I, II, III.

Hofmann, Joseph E. *The History of Mathematics*. Translated by F. Gaynor and H. O. Midonich. Philosophical Library, Inc., New York, 1957.

Hogben, Lancelot. *Mathematics for the Millions*, 4th ed. W. W. Norton & Company, Inc., New York, 1968.

Hooper, Alfred. *Makers of Mathematics*. Random House, Inc., New York, 1948.

Huntley, H. E. *The Divine Proportion: A Study in Mathematical Beauty*. Dover Publications, Inc., New York, 1970.

Kline, Morris. *Mathematics, A Cultural Approach*. Addison-Wesley Publishing Co., Inc., Reading, Mass., 1962.

Kline, Morris, ed. *Mathematics in the Modern World*. W. H. Freeman and Company, Publishers, San Francisco, 1969.

Loomis, Elisha Scott. *Mistakes in Geometric Proofs*. National Council of Teachers of Mathematics, Washington, D.C., 1968.

Neugebauer, O. *The Exact Science in Antiquity*, 2d ed. Brown University Press, Providence, R.I., 1957.

Newman, James R. *The World of Mathematics*. 4 vols. Simon & Schuster, Inc., New York, 1956.

Van der Waerden, B. L. *Science Awakening*. Translated by A. Dresden. Oxford University Press, New York, 1961.

Zippin, Leo. *Uses of Infinity*. Random House, Inc., New York, 1962.

Part 3

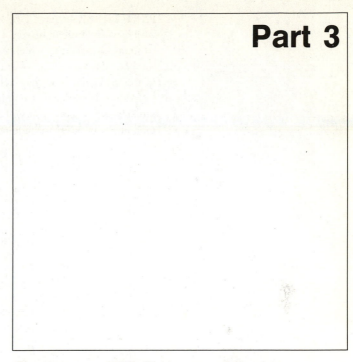

Some Things New

PART THREE In Part Two we took a new look at some mathematics which were familiar to you. In Part Three we are going to introduce you to some mathematics which may be new. As we go along, we will again see how abstractions grow out of real problems (more involved problems in this part) and how they help to solve them.

Although we haven't emphasized the learning of techniques and we aren't going to now, you will find it necessary to learn some new techniques in order to understand more thoroughly what is going on.

Chapter 5

Logic and Sets

5.1 Introduction

It should be pretty clear by now that the fundamental tools in mathematics are words. We use words to describe nature; we use words to form our abstractions; we even use words to think about other words. Therefore, it isn't very surprising that a mathematical theory of words and phrases has been developed. This branch of mathematics—or logic —is often called *the sentential calculus*; the word *sentential* stems from the same root as the word *sentence*. In this logic we will develop a calculus—that is, a set of rules of manipulation, not the calculus of the physicist and engineer—for mathematical sentences, which will allow us to transform these sentences in ways that will preserve their truth values. Once we have found this calculus, we will put it to work for us in developing a theory for mathematical sets that many regard as the basis of all mathematical theory.

5.2 Basic Statements and Truth Values

We will start, as we must in any branch of mathematics, with our undefined terms. To begin with, we need the notion of *true* and *false*, which we shall abbreviate T and F respectively. We also need the idea of mathematical *statements* (we shall designate them by the letters p, q, r, s, . . .) which will be the variables of our calculus. The only property that we shall insist upon is that a mathematical statement must have *exactly one* of two possible truth values, T or F.

Observe that what we have set up is a system which is an abstraction of the usual statements we make in mathematics. Some mathematical statements have only a single truth value. For example "$2 + 2 = 5$" is always false, while "Every equilateral triangle is also equiangular" is always true. On the other hand, some mathematical statements have varying truth values; $x^2 + 3 = 7$, for example, is true only when $x = 2$ or $x = -2$, and false for all other values of x, but in any case is never anything but T or F. What we have done is to eliminate the undecided (or undecidable) statements of mathematics from our calculus. We are insisting on what is often called a two-valued logic.

We have also eliminated from consideration statements whose truth value is a matter of opinion. The statement "Dwight David Eisenhower was a poor president of the United States" will have different truth values for different people even under the same set of circumstances. Similarly, it is unlikely that it would be possible to get a general agreement about any statement which includes a subjective value judgment. Thus, it is an important part of the undefined basis of our logic that there exists some kind of machinery for determining the truth values of our primitive statements.

We begin our study of logic by considering *connectives*. That is, we are going to "fasten together" two or more statements in order to obtain a new statement. Then we will investigate the truth value of these new statements under varying conditions of the truth values of the original statements that went into the new one.

Of all connectives used in English, the word *and* is one of the most often encountered. Suppose we begin our study of the truth values of the new sentences we construct using *and* by considering an example: *Bill is six feet tall and Deborah is five feet tall.* Two components have gone into making up the sentence:

Bill is six feet tall

is the first, and

Deborah is five feet tall

is the second, and they have been "glued" together with the conjunction *and*. The question now arises, when is the *entire* sentence "Bill is . . . and Deborah . . . tall" true and when is it false? Clearly this will depend on the truth values of each of the components. Now there are four possible pairs of truth values for the components.

1. Both are *true*.
2. The first is *true* and the second is *false*.
3. The first is *false* and the second is *true*.
4. Both are *false*.

Observe that we could abbreviate these possibilities as in Table 5.1.

Table 5.1

1.	*T*	*T*
2.	*T*	*F*
3.	*F*	*T*
4.	*F*	*F*

These represent the total possible truth input into the sentence, and we will call this table the set of *initial conditions* for our sentence.

Since the word *and* in English means essentially the same as *both*, we see that the entire sentence will be true only when the components are *both* true, that is, under initial condition 1. Moreover, since the sentence is a statement of our logic, it must be false where it is not true; hence it is false under initial conditions 2, 3, and 4. We can compile the above results into Table 5.2.

Table 5.2

TT	*T*
TF	*F*
FT	*F*
FF	*F*

Now, let's consider the whole matter in the abstract. If we have two abstract statements, which we will denote by p and q, and if we take the symbol $\wedge$ to mean *and*, then the set of symbols $p \wedge q$ represents the new statement "p and q". Furthermore, the same initial conditions hold, that is:

1. p and q are both *true*.
2. p is *true* and q is *false*.
3. p is *false* and q is *true*.
4. p and q are both *false*.

We represent this by exactly the same table of initial conditions. In addition, since our set of symbols $p \wedge q$ is abstracted from the

English, and since we wish our abstractions to be reflections of reality (just as we do in algebra and geometry), we use the display shown in Table 5.3, called a *truth table*:

Table 5.3

p	q	$p \wedge q$
T	T	T
T	F	F
F	T	F
F	F	F

This we take as the *definition* of the truth values for the new statement $p \wedge q$ under the varying truth possibilities of p and of q.

The next connective of importance is "or." Here we run into trouble; the English is ambiguous. If we say, "I have enough money to buy you a beer or take you to a movie," the implication is clear that there are insufficient funds for both. On the other hand if we say "I get higher grades than Fred or Harry," there is no reason to suspect that I am not getting higher grades than both. So we see that "or" exists in ordinary English in two forms, the exclusive form (which is more usual) and the inclusive. Lawyers get around this ambiguity with the ugly "and/or" whenever they want the inclusive form.

Mathematicians, perhaps because they eschew the unaesthetic, take a different way out. Unless it is crystal clear from the context that the "or" is intended to be exclusive, mathematicians *always* use "or" in the *inclusive* sense. If we wish to indicate the exclusive meaning, "but not both" must be added. Thus we can say, correctly, that "An algebraic statement contains numbers or symbols which stand for numbers." We also say "A number is positive or negative" without adding "but not both" since it is perfectly obvious that the exclusive sense is meant here. But we could not say "A trapezoid which is not a parallelogram has one pair or the other of opposite sides parallel," unless we added "but not both."

What we are leading up to is this: For the mathematician a statement of the form "I have enough money to buy you a beer or take you to a movie," which can be re-expressed as "I have enough money to buy you a beer or I have enough money to take you to a movie" is true when *both* components are true, contrary to more usual English usage. Clearly, the compound statement is also true when just *one* of the components is true, and is false only when *both* are false.

Again, in the abstract situation, things are the same. We take the symbol $\vee$ for *or* and investigate the truth values for $p \vee q$. We note that we have exactly the same initial conditions as for $p \wedge q$, so only the last column changes. In this case, $p \vee q$ is true when either or both

p and q are true, and false only when they are both false. Thus, our truth table looks like Table 5.4.

Table 5.4

p	q	$p \vee q$
T	T	T
T	F	T
F	T	T
F	F	F

Thus, as before, we *define* this tabulation to be the truth values of $p \vee q$ under the respective initial conditions.

The connective "$\wedge$" is called the *conjunction*; you may find it easier to remember if you notice that it is like the upper-case form of the first letter in And, except that it has no crossbar. The "$\vee$" is called the *disjunction*, and I don't have a mnemonic device to help you.

Our last connective is *implication*. Consider the statement "If a triangle is isosceles, then it has equal base angles." We can decompose it into its component parts:

A triangle is isosceles

and

It (a triangle) has equal base angles

In our implication, these components are connected by "if . . . then . . ." In this case we will consider the entire sentence to be true if there is a valid deductive argument with the first component as *hypothesis* and the second component as the *conclusion*. Thus, the above statement is true, while "If a triangle is equilateral, then one of its angles is larger than the others" is false.

Unfortunately, since implication is a more complex connective than *and* or *or* the abstract formulation is also a bit more complicated. We will symbolically denote it by $\rightarrow$, so that the expression $p \rightarrow q$ means "p implies q" or "if p then q." In the form $p \rightarrow q$, p is the *hypothesis* and q is the *conclusion*.

To form the truth table, we must take our cue from Chapter 1. We know that a true hypothesis can lead to a true conclusion, so to the initial condition TT, corresponding to a true hypothesis and a true conclusion, we assign the value t to $p \rightarrow q$. We also know that a true hypothesis cannot validly lead to a false conclusion, so we assign F to the initial condition TF. Similarly, we know that false hypotheses *can* lead to a true conclusion, so we assign to FT the truth value T by definition, and also a false hypothesis can lead to a false conclusion, so the value T is similarly assigned to FF.

This gives us a defining truth table for $p \to q$ which looks like Table 5.5.

Table 5.5

p	q	$p \to q$
T	T	T
T	F	F
F	T	T
F	F	T

Finally we take up *negation*. This is not a connective, since it operates on only a single statement. Clearly, in ordinary English, if a statement is true, its negation is false, and *vice versa*. We will denote negation by $\sim$; that is, "not p" is written $\sim p$. Since p is the only variable and since it can assume only the values T or F, we have only two initial conditions, and our truth table for negation looks like Table 5.6.

Table 5.6

p	$\sim p$
T	F
F	T

We summarize all the truth tables in one large one (Table 5.7) for convenience:

Table 5.7

p	q	$p \wedge q$	$p \vee q$	$p \to q$	$\sim p$
T	T	T	T	T	F
T	F	F	T	F	
F	T	F	T	T	T
F	F	F	F	T	

We emphasize that these tables are the *defining* truth values for the given symbols under the varying input conditions. However, observe that these definitions are not arbitrary, but have been selected to reflect the reality which gave rise to the particular abstractions.

EXERCISES 5.2

1. Which of the following are sentences which are admissible into the sentential calculus?
   (a) Jack is taller than Jill.
   (b) Mathematics is useful.
   (c) The poetry of Edgar Allan Poe is dreadful.
   (d) $6 + 7 = 12$.
   (e) I need 8 more credit hours to graduate.

(f) .101001000100001 · · · is a rational number.

(g) .101001000100001 · · · is a transcendental number.

2. Can the statement "Mary and Joan are both honor students" be recast into the form _____ and _____? If so, do it; if not, explain. Under what conditions will it be a true statement if it is to be taken in its entirety?

3. Answer the same questions as in Exercise 2 for the sentence "Purple is blue and red." What is the difference here?

4. Answer the same questions as in Exercise 2 for the sentence "Two and seven are nine."

5. Under exactly what conditions will the following sentences be true considered *as a whole*? Be very specific in your answers.
   (a) Texas is the largest state and California is the most populous.
   (b) Frank and John stepped out for a beer.
   (c) Frank stepped out for a beer and a paper.
   (d) Frank couldn't buy a beer and a paper.
   (e) Frank could buy neither a beer nor a paper.

6. Wherever possible write the following statements as "_____ or _____." See if you can determine when each is true either in the ordinary sense or in the mathematical sense. Be very explicit.
   (a) On maps the oceans are usually colored blue or green.
   (b) Neither snow, nor rain, nor heat, nor gloom of night stays these couriers from the swift completion of their appointed rounds.
   (c) The dinner includes salad and a vegetable or two vegetables.
   (d) Jeanne has one or two A's this semester.

7. Analyze the exclusive *or* as we did the inclusive *or*, and construct a truth table for it.

8. Recast the following so they fit into the "If _____ then _____" form, rewriting if necessary. Identify the components and decide, where possible, whether the entire statement is true.
   (a) When it rains, I get my feet wet.
   (b) It must follow, as the night the day . . . .
   (c) Whenever the moon is blue, $2 + 2 = 5$.

9. Explain what is meant by "from a false hypothesis you can prove anything."

10. Make a copy of the summarized truth table for reference in the subsequent sections.

5.3 Properties of Connectives

Now that we have our defining truth tables, we can study the properties of the connectives represented by them. However, before we do that, we need to know what is meant by equal statements. We shall say that two statements are *equal* (or *equivalent*) if they have the same truth table, that is, if for every initial condition, the corresponding truth value of each is the same. This means that if we actually substitute real statements into our equal logical statements, the truth values will remain unchanged from one equal form to another.

Suppose we start with $p \wedge q$ and $q \wedge p$. The statement $p \wedge q$ has as its truth table 5.8.

Table 5.8

p	q	$p \wedge q$
T	T	T
T	F	F
F	T	F
F	F	F

We now start to make the truth table for $q \wedge p$. Since we wish to compare the truth tables initial condition by initial condition, we set the initial conditions up in the same way. Thus we start with a work sheet which looks like Table 5.9.

Table 5.9

p	q	$q \wedge p$
T	T	
T	F	
F	T	
F	F	

Our job is to fill in the empty column with T's and F's that will be the truth values for $q \wedge p$.

The next step is to fill in the appropriate truth values under the p and under the q, so after our second step we have the array in Table 5.10.

Table 5.10

p	q	q	$\wedge$	p
T	T	T		T
T	F	F		T
F	T	T		F
F	F	F		F

Now to finish, we fill in the column under the $\wedge$, being sure to use the *new* columns of initial conditions under the p and the q. Thus in the first row we have a TT which, when joined by a $\wedge$, gives T. In the second row we have an *FT* which gives *F* under ∧. Similarly, *TF* of the third row gives *F*, while the *FF* of the last row also gives *F*. In every case, we have gotten our assigned truth value from the defining truth table, being sure to use the correct set of initial conditions in the *correct order*.

We put this all together in Table 5.11, where the numbers under the columns indicate the order of the appropriate steps.

Table 5.11

p	q	$q \wedge p$
T	*T*	*T T T*
T	*F*	*F F T*
F	*T*	*T F F*
F	*F*	*F F F*
		1 2 1

Notice that the final column (column 2) gives us the truth values for $q \wedge p$, which are identical with the truth values for $p \wedge q$, identical in values and in order. Thus we have that $p \wedge q = q \wedge p$ under our definition of equality, which tells us that the operation represented by ∧ is *commutative*.

Next, suppose we consider $p \rightarrow q$ and ask whether it is commutative; that is, is $p \rightarrow q = q \rightarrow p$ a true statement? We know that the truth for $p \rightarrow q$ is by definition Table 5.12.

Table 5.12

p	q	$p \rightarrow q$
T	*T*	*T*
T	*F*	*F*
F	*T*	*T*
F	*F*	*T*

We next consider $q \rightarrow p$. Following the method we just used, we arrive at Table 5.13 for the first step.

Table 5.13

p	q	$q \rightarrow p$
T	*T*	*T T*
T	*F*	*F T*
F	*T*	*T F*
F	*F*	*F F*

Now, working under the → and using the defining table for implication, we have that, to the initial condition *TT* is assigned a *T*; to *FT in that order* is assigned a *T*; to *TF in that order* is assigned an *F*; and to *FF* is assigned a *T*. This gives us Table 5.14.

Table 5.14

p	q	$q \rightarrow p$
T	T	$T \; T \; T$
T	F	$F \; T \; T$
F	T	$T \; F \; F$
F	F	$F \; T \; F$
		$1 \; 2 \; 1$

Since column 2, the final column, does not agree with the truth table for $p \rightarrow q$ (rows 2 and 3 are different), we have shown that $p \rightarrow q$ and $q \rightarrow p$ have different truth values under similar initial conditions, so that $\rightarrow$ is *not commutative*. For example, "If a figure is a quadrilateral, then it is a square" is logically false, since there are many figures of four sides which are *not* squares. On the other hand "If a figure is a square, it is a quadrilateral" is logically true.

Now this is, surely, a result we could expect. Just because a particular conclusion q follows from a hypotheses p does not necessarily mean that p follows logically from q, as the example shows. Implication is *not*, in general, reversible. Thus, in a limited way, we have already had a chance to check the results of our abstractions against nature.

A glance at the algebra axioms in Chapter 3 suggests that the next property to investigate for our three connectives is associativity. Associativity for $\vee$ would look like this; $(p \vee q) \vee r = p \vee (q \vee r)$. In order to test this, we have to worry about each of three variables *independently*, assuming the two truth values T or F. To do this we need eight sets of initial conditions in order to cover all possible ways that p, q, and r can each be true or be false. We arrange them as in Table 5.15.

Table 5.15

p	q	r
T	T	T
T	T	F
T	F	T
T	F	F
F	T	T
F	T	F
F	F	T
F	F	F

Pause here to note the use of parentheses in our statement. We must treat what is inside them as a *single entity*. That means we must find the truth value of the expression *inside* first; and when there are several, all nested one inside the other, we start at the *innermost* set and work our way out.

To save time in evaluation, we will treat both expressions side by side, using the same set of initial conditions for both. (See Table 5.16.)

Table 5.16

p q r	(p ∨ q) ∨ r	p ∨ (q ∨ r)
T T T	T T T	T T T
T T F	T T T	T T F
T F T	T T F	F T T
T F F	T T F	F F F
F T T	F T T	T T T
F T F	F T T	T T F
F F T	F F F	F T T
F F F	F F F	F F F
	1 2 1	1 2 1

Here we have just copied down the truth values for p and q in the first expression, and q and r in the second (columns numbered 1), and then found the truth values for the expressions inside the parentheses marked column 2 in each case).

We now have to determine the relation between the truth values we have found inside the parentheses and the balance of the expression. Continuing, we repeat the values just found for the statements in the parentheses and also write down the values for r and p in the respective cases; this gives us column 3. We now have, in each expression, two columns of T's and F's (2 and 3) joined by a ∨. We know from its definition that the ∨ assigns the value T to all initial conditions except FF, so our final column (4) assigns T to all but the last row in each case (see Table 5.17).

Table 5.17

p q r	(p ∨ q) ∨ r	p ∨ (q ∨ r)
T T T	T T T	T T T
T T F	T T F	T T T
T F T	T T T	T T T
T F F	T T F	T T F
F T T	T T T	F T T
F T F	T T F	F T T
F F T	F T T	F T T
F F F	F F F	F F F
	2 4 3	3 4 2

Since the final columns (4) are the same, we see that the expressions are equal and that we have an *associative* law. Thus, we can drop the parentheses in the future; and furthermore, from the truth table we note that $p ∨ q ∨ r$ is true when even just *one* of its components is true, and false only when *all three* are false.

We can next ask "How do our connectives interact among themselves?" Do we have any *distributive* laws? For example, is $\vee$ distributive with respect to $\wedge$; that is, is $p \vee (q \wedge r) = (p \vee q) \wedge (p \vee r)$? We proceed as before, using the same set of initial conditions for each expression and starting with the parentheses.

Table 5.18

$p\ q\ r$	$(p \vee q) \wedge (p \vee r)$		$p \vee (q \wedge r)$
$T\ T\ T$	$T\ T\ T$	$T\ T\ T$	$T\ T\ T$
$T\ T\ F$	$T\ T\ T$	$T\ T\ F$	$T\ F\ F$
$T\ F\ T$	$T\ T\ F$	$T\ T\ T$	$F\ F\ T$
$T\ F\ F$	$T\ T\ F$	$T\ T\ F$	$F\ F\ F$
$F\ T\ T$	$F\ T\ T$	$F\ T\ T$	$T\ T\ T$
$F\ T\ F$	$F\ T\ T$	$F\ F\ F$	$T\ F\ F$
$F\ F\ T$	$F\ F\ F$	$F\ T\ T$	$F\ F\ T$
$F\ F\ F$	$F\ F\ F$	$F\ F\ F$	$F\ F\ F$
	$1\ 2\ 1$	$1\ 2\ 1$	$1\ 2\ 1$

In each case we have written in the values for the appropriate variables (columns 1) and determined the values for each of the parentheses (columns 2).

We finish by finding the correct values under the $\wedge$ in the first expression, and by putting in the values for p and combining with column 2 under the $\vee$ in the second, which gives the column labeled 4 in each case (see Table 5.19).

Table 5.19

$p\ q\ r$	$(p \vee q) \wedge (p \vee r)$			$p \vee (q \wedge r)$		
$T\ T\ T$	T	T	T	T	T	T
$T\ T\ F$	T	T	T	T	T	F
$T\ F\ T$	T	T	T	T	T	F
$T\ F\ F$	T	T	T	T	T	F
$F\ T\ T$	T	T	T	F	T	T
$F\ T\ F$	T	F	F	F	F	F
$F\ F\ T$	F	F	T	F	F	F
$F\ F\ F$	F	F	F	F	F	F
	2	4	2	3	4	2

Thus we have at least one distributive law, namely, that $\vee$ is distributive with respect to $\wedge$. As it turns out, there are *two* distributive laws of interest in logic, the other being $p \wedge (q \vee r) = (p \wedge q) \vee (p \wedge r)$.

We formulate a chart indicating those properties we will use later on.

$\wedge$: commutative, associative, distributive with respect to $\vee$

$\vee$: commutative, associative, distributive with respect to $\wedge$

$\rightarrow$: not commutative.

EXERCISES 5.3

1. Use the technique of the text to show that $\vee$ is commutative.

2. Does it make any sense to ask if $\sim$ is commutative? If so, is it? Explain.

3. Do we have closure property (see Chapter 2) for our three connectives?

4. When we have one variable we need two initial conditions; when we have two variables we need four initial conditions; when we have three variables we need eight initial conditions. How many initial conditions do we need for four, five, six, or n variables? Write down the list of initial conditions for five and six variables. What does this suggest about the efficacy of truth tables for large numbers of variables?

5. Show that $\wedge$ is associative. Describe in words the significance of the truth table you found.

6. Is $\rightarrow$ associative? Construct a table and explain the result.

7. Show that $\wedge$ is distributive with respect to $\vee$; that is, show that $p \wedge (q \vee r) = (p \vee q) \wedge (p \vee r)$.

8. Formulate all other distributive laws using $\rightarrow$ with $\vee$ and $\wedge$. Which are correct laws?

9. The connective *if and only if*, denoted by $\leftrightarrow$, has as its defining truth table (with the initial conditions in the usual order) *T, F, F, T.*
   (a) Show that $p \leftrightarrow q = (p \rightarrow q) \wedge (q \rightarrow p)$.
   (b) Explain in words the meaning of $p \leftrightarrow q$, using (a) above to help you.
   (c) Go back to the discussion in Chapter 2 concerning repeating decimals and rational numbers. What does this discussion have to do with part (b)?
   (d) Determine which properties of commutativity, associativity, etc., $\leftrightarrow$ has.

5.4 Negation and the Connectives

Since negation is not a connective, it makes little sense to ask the same questions about it as we have just asked about the connectives. However, we can ask how the negation process interacts with each of the other connectives.

We first take up the question of what negation does to itself. Not surprisingly,

$$\sim(\sim p) = p$$

To see this, we set up our initial conditions (only two are needed), and first find the truth value of the expression inside the parentheses and then the value of the whole expression.

Table 5.20

p	$\sim(\sim p)$
T	T F
F	F T
	2 1

Column 1 is from the table for $\sim p$, as is column 2, where we notice that $\sim$ assigns T to F and F to T. Since our last column 2 agrees with p, we have established our result.

It might be tempting, at this stage, to suppose that, just because the negation looks, sounds, and acts (in this one case) somewhat like the negative sign of arithmetic, it behaves like it. For example, you might conjecture that $\sim(p \lor q) = (\sim p) \lor (\sim q)$. Let's try it. Only four sets of initial conditions are needed, and we start by working inside parentheses (columns 1) and then finish up in the columns numbered 2.

Table 5.21

p	q	$\sim(p \lor q)$		$(\sim p) \lor (\sim q)$		
T	T	F	T	F	F	F
T	F	F	T	F	T	T
F	T	F	T	T	T	F
F	F	T	F	T	T	T
		2	1	1	2	1

Clearly these are *not* the same, so we cannot just "multiply in" the negation sign.

A clue to where the correct statement equivalent to $(\sim p) \lor (\sim q)$ comes from can be discovered by looking carefully at its column 2. Some thought and a look at our various defining truth tables show us that we would get the same thing if we took the negation of $p \land q$; that is, $\sim(p \land q) = (\sim p) \lor (\sim q)$. A similar discussion shows that $\sim(p \lor q) = (\sim p \land (\sim q)$, and these are called De Morgan's rules, after the nineteenth-century British logician, Augustus de Morgan.

If we look at the relation between $\sim$ and $\rightarrow$, we find that it is not possible that $\sim(p \rightarrow q) = (\sim p) \rightarrow (\sim q)$ since the left side has one T

and three F's while the right side has one F and three T's. (A similar argument would also have shown that $\sim(p \lor q)$ is not $(\sim p) \lor (\sim q)$.) But let's see what happens to $(\sim p) \to (\sim q)$:

Table 5.22

p	q	$(\sim p) \to (\sim q)$		
T	T	F	T	F
T	F	F	T	T
F	T	T	F	F
F	F	T	T	T
		1	2	1

Now column 2 looks familiar. If we go back to the tables in our earlier discussion about the commutativity of $\to$, we see that $(\sim p) \to (\sim q) = q \to p$. Similar considerations show that $(\sim q) \to (\sim p) = p \to q$. This last equality is called the *law of contraposition*, and the statement $(\sim q) \to (\sim p)$ is called the *contrapositive* of $p \to q$. Notice that

$$(\sim q) \to (\sim p) = p \to q$$

This is very important to mathematics, as it provides the logical basis for a style of proof called *proof by contradiction*, or *reductio ad absurdum* (reduced to an absurdity), or *proof by indirect method*. We used this when we proved $\sqrt{2}$ irrational. Recall in that proof by contradiction, we assume that the conclusion is false, even though all of the hypotheses are true. Then we prove, by a valid deductive argument, that one of the hypotheses must be false, and this is where the contradiction comes in. That is, a false conclusion implies a false hypothesis, a fact we have noted earlier.

As another example, we shall prove that: Given L_1 parallel to L_2, then (referring to Figure 5.1) $\measuredangle\, a = \measuredangle\, b$, where we shall use Euclid's fifth postulate.

Suppose $\measuredangle\, a$ is *not* equal to $\measuredangle\, b$. Then one of them, say $\measuredangle\, a$, is greater than the other, $\measuredangle\, b$.

But $\measuredangle\, a + \measuredangle\, c = 180°$ and $\measuredangle\, b + \measuredangle\, d = 180°$; since $\measuredangle\, a + \measuredangle\, c$ and $\measuredangle\, b + \measuredangle\, d$ are straight angles.

Figure 5.1

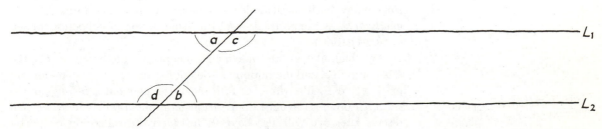

Hence, $\angle\, c$ is less than $\angle\, d$.

And so, $\angle\, b + \angle\, c$ is less than 180°

Thus L_1 meets L_2 to the right, according to the fifth postulate.

Similarly, if $\angle\, a$ is less than $\angle\, b$, L_1 meets L_2 to the left.

Both of these alternatives are absurd, since L_1 was *given to be parallel* to L_2.

Thus we are left with the only remaining possibility, namely, $\angle\, a = \angle\, b$.

EXERCISES 5.4

1. Verify De Morgan's rules directly, using truth tables.

2. Verify the law of contraposition directly.

3. Referring to Exercise 5.3.9, investigate, in two different ways, the relation between negation and the connective "if and only if." One of the two tells you something about a relation between "if and only if" and the exclusive "or."

4. Prove that the three angles of an equilateral triangle are all equal, using the indirect method. Use the theorem that if two angles of a triangle are unequal, the larger is opposite the longer side.

5. In the example of indirect proof given in the text, a small change in the last statements gives a direct illustration of $\sim q \rightarrow \sim p$.
   (a) Restate the theorem so it has the form $\sim q \rightarrow \sim p$.
   (b) Change the proof to illustrate the indirect method, as indicated above.

5.5 Sets

Another of the basic tools of mathematics is the mathematical set, since many mathematical ideas can be framed in terms of sets. A *mathematical set* is the abstraction of the usual notion of collection, or set, or aggregation.

We shall take, as our undefined concepts in our theory of sets, the sets themselves and the notion of "belonging to" or "being an element of" a set. We will write $a \in S$ to mean the element *a belongs to the set S*. In order to avoid rather serious complications, which we will discuss later, we will also have to insist on the following property:

No set can be an element of itself. (For example, this prohibits us from forming the set of all sets, since it is a *set* which would contain itself (it is a *set*) as an element.)

One of the characteristic properties of a set is that it has a defining mathematical statement. That is, associated with each set is a statement which is *true* for *precisely* those objects which are elements of the set, and is *false* for all *other* objects. This statement takes essentially two different forms; one is in the nature of a *list*, and the other is *descriptive* of the properties of the elements of the set.

When we are dealing with the *list*, our defining statement is really a shorthand way of saying "The object is one of the following items: . . . ." For example, the set of integers from -1 to 3, inclusive, could be denoted by $\{-1, 0, 1, 2, 3\}$. Note first the curly brackets (or braces), which are always used in connection with sets. Also observe that, in this case, because there are few elements in the set, we may be able to give a list of all the elements.

However, suppose we wanted the set of integers from 3 to 109, inclusive. We *could* go ahead and list all of the elements, but the result would be clumsy and the work would be tiresome, to say the least. There is a better way. We can indicate the list without actually writing everything down. Thus, our set could be written as $\{3, 4, 5, \ldots, 109\}$. We start as if we were going to write out the entire list, but as soon as we have enough entries to make it absolutely clear how we are to proceed (a minimum of three), we put in the three dots. These mean that we are to continue as before, and then the 109 tells us when to stop.

Often we are faced with the prospect of a list which is infinite in extent. Suppose we wished to form the set of all even natural numbers. Since this list is endless (literally, it has no upper end), we couldn't write it all down, even if we wanted to. What we *can* do, though, is to indicate its membership as before, writing our set as $\{2, 4, 6, \ldots\}$. Notice, in this case, there is no stopper; the three dots say "Just keep going." Sometimes, the three dots come at the beginning of the list; thus the set of negative even integers could be written $\{\ldots, -6, -4, -2\}$, and the set of all even integers as $\{\ldots, -4, -2, 0, 2, \ldots\}$.

The other type of defining statement is the *descriptive* type. For example, we could call for the set of all counting numbers, or we could be more formal and write the same set as:

$$\{x: x \text{ is a counting number}\}$$

The notation is read "the set of all x such that x is a counting number." It is customary, when forming sets, to label the objects in some way with a variable (such as x in the above example) and use a mathematical statement. That means we could use as a label essentially any statement which has only two possible truth values, T or F. Usually, one has a particular collection of objects in mind, and the difficulty comes in selecting a correct mathematical statement so that the resulting set is what is wanted, neither too small nor too large. This statement is called a *defining statement* for the set.

EXERCISES 5.5

1. Describe the following sets in words.
    (a) $\{1, 2, 3, \ldots, 100\}$
    (b) $\{1, 2, 3, \ldots\}$
    (c) $\{0, 3, 6, 9 \ldots\}$
    (d) $\{\ldots, -3, 0, 3, \ldots\}$
    (e) $\{1, 4, 9, 16, 25, 36, 49, 64, 81, 100\}$
    (f) $\{1, \sqrt{2}, \sqrt{3}, 2, \sqrt{5}, \sqrt{6}, \sqrt{7}, 2\sqrt{2}, 3\}$

2. Write the following sets, using indicated lists for the defining statements.
    (a) The set of integers between 6 and 13 inclusive.
    (b) The set of integers between 6 and 8.
    (c) The set of negative integers.
    (d) The set of integers evenly divisible by 3.
    (e) The set of all numbers which are neither positive nor negative.
    (f) The set of integers between 1 and 2 exclusive.
    (g) The set of odd positive integers.

3. Describe in words or in list form the following sets.
    (a) $\{x : x \text{ is positive}\}$
    (b) $\{x : -6 \le x \le 2\}$
    (c) $\{y : y = \sqrt{1 - x^2}, x \text{ is a real number between } -1 \text{ and } 1\}$
    (d) $\{z : z \text{ is a line parallel to a horizontal line}\}$
    (e) $\{d : d \text{ is a repeating decimal}\}$
    (f) $\{s : \sqrt{s} \text{ is an integer}, s \le 144\}$

5.6 Properties of Sets

Following the pattern we started with statements, we shall now examine certain operations on sets, and how these operations interact with each other. Our task will be easier than before, because we can use the defining statements for sets and apply to them our earlier results from logic.

We start with defining equality for sets. Two sets are *equal* if all the elements of one are elements of the other, and *vice versa*. More formally, the set $A =$ the set B if, whenever $a \in A$, then also $a \in B$ *and* whenever $b \in B$, then it must be that $b \in A$. Observe that the order of writing down the objects is unimportant; the *collection* of objects is the same. Thus, if

$$A = \{1, 3, 7, 5\} \qquad \text{and} \qquad B = \{3, 5, 7, 1\}$$

we have $A = B$.

In terms of defining statements, this means that if p is the defining statement for A and if q is the defining statement for B, then $A = B$ is the same as saying that the statement p and the statement q are true for precisely the same collection of objects and are also false for the same set of objects. For example, let A be the set of all rational numbers and B be the set of all repeating decimals. Thus p is the statement "x is a rational number" and q is "x is a repeating decimal," and we know, from our work in Chapter 2 that p and q are true for the same objects and false for the same objects; this is the same as saying $A = B$.

An important relation between sets is the idea of a *subset*. We say that A is a subset of B (in symbols, written as $A \subset B$ or $B \supset A$) if every element of A is also an element of B. For their defining statements, this is the same as saying that q, the defining statement for B, must be true whenever p, the definer for A, is true. Note that q *may* also be true when p is not.

In particular, suppose we have a set whose defining statement is always false; for example, $\{x : x \text{ is different from } x\}$. Such a set has *no* elements in it, and is called the *empty set*. Observe that the empty set is *still a set*, even though there are no objects in it, just as an empty milk carton is still a milk carton.

Suppose we denote one of the defining statements for the empty set by f; then f is always false. Now, if q is *any* statement, q is always true whenever f is true (since f is never true this is an easy requirement for q to fulfill), so that no matter *what* set q defines, the empty set will be a subset of it. That is, the *empty set is a subset of every set*. The empty set is also called the *null* set, and has its own notation, $\emptyset$ or $\{\}$. (Other texts also use N and E.)

The two principal operations for sets can easily be described in terms of the connectives of their defining statements. If we again take p and q as the defining statements for A and B, respectively, then $p \wedge q$ is a defining statement for the set of elements common to both

A and *B*. This new set is called the *intersection* of *A* and *B* and is written *A* ∩ *B*.

Similarly, the new set which is formed of all of the elements *either* in *A* or in *B* (or in *both*, of course) has as one of its defining statements $p \lor q$ and is called the *union* of *A* and *B*, written *A* ∪ *B*. To see how these work, just refer to the defining truth tables and you will see that $p \land q$ is true only when both *p* and *q* are true, while $p \lor q$ is true when either *p* or *q* (or both) are true.

If we picture *A* as a circular area, and *B* as a triangular area, then *A* ∩ *B* looks like the shaded area in Figure 5.2,

Figure 5.2

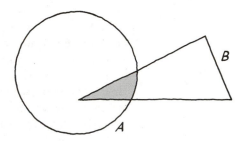

while *A* ∪ *B* looks like the shaded area in Figure 5.3.

Figure 5.3

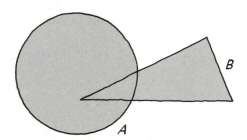

Notice that, since *A* and *B* are mathematical sets, it follows that *A* ∪ *B* and *A* ∩ *B* are also mathematical sets, so we have the closure property for both union and intersection. However, if we did not have the empty set, this would not be so, because how could we have formed the intersection of {1, 2} and {3, 4}?

Once we have established the operations of intersection and union on sets in terms of their defining statements, the properties of these operations follow at once from the properties of their defining statements. Thus, both intersection (∩) and union (∪) are *commutative* and *associative*, for if *p*, *q*, and *r* are defining statements for *A*, *B*, and *C*, respectively, we have that $p \land q = q \land p$, and $p \lor q = q \lor p$; and also, we have

$$(p \wedge q) \wedge r = p \wedge (q \wedge r)$$

and

$$(p \vee q) \vee r = p \vee (q \vee r)$$

If A and B are as above, and C is a trapezoidal area, then $A \cap B \cap C$ is the shaded area in Figure 5.4,

Figure 5.4

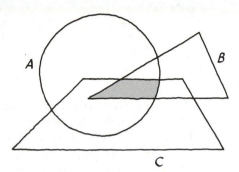

and $A \cup B \cup C$ is the shaded area in Figure 5.5.

Figure 5.5

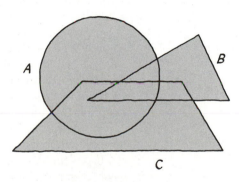

We also have two distributive laws for sets, namely, $A \cap (B \cup C) = (A \cap B) \cup (A \cap C)$, which looks like the shaded area in Figure 5.6,

Figure 5.6

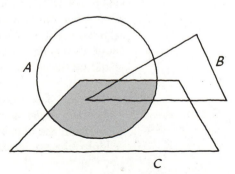

and $A \cup (B \cap C) = (A \cup B) \cap (A \cup C)$, which looks like the shaded area in Figure 5.7

Figure 5.7

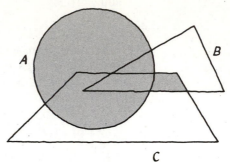

Of course, both of these laws follow from the distributive laws for $\lor$ and $\land$ in logic.

The laws of negation operate in very much the same manner, except that we have to exercise a little care since we can't allow a set to be a member of itself. Again, suppose A is a set with defining statement p. Since $\sim p$ is true for every object in the universe for which p is false, if we tried to take $\sim p$ as the defining statement of a set, the result would be an inadmissible set, since it could very easily include itself as an element. To avoid this possibility, we must limit ourselves as follows: When we are dealing with sets in regard to a particular problem, we set up a universe *with regard to that problem* consisting of everything conceivable connected with that problem. We call this set the *universal* set, denoted by U, and then proceed under the assumption that all of our sets are subsets of the universal set.

To see this, let's look at an example. Suppose we were interested in a problem involving the lines and circles of Euclidean plane geometry. If p is the statement "x is a horizontal line," then $\sim p$ is the statement "x is not a horizontal line." Thus if we try to form the set B, whose defining statement is $\sim p$, we get that B is a set which not only includes lines which are not horizontal, but also *all objects of any kind which are not horizontal lines*. This includes the *set* of objects which are not horizontal lines, since the set is, itself, an object, which is not a horizontal line. That is, B is a member of itself, which we cannot allow. In this case a reasonable universal set for our lines and circles would be the set of all lines and circles in the Euclidean plane of our problem.

We can now form the set whose defining statement is $\sim p$ and call it the *complement* of A (denoted by $\sim A$). Thus $\sim A$ is the set of all elements in the universal set which are *not* in A. The complement of A looks like the shaded area in Figure 5.8,

Figure 5.8

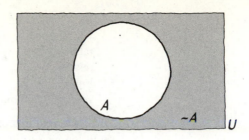

where the big rectangle represents the universal set and the disk represents A.

It is now easy to state De Morgan's rules for sets. Since $\sim(p \vee q) = (\sim p) \wedge (\sim q)$, we have that $\sim(A \cup B) = (\sim A) \cap (\sim B)$, and similarly, $\sim(A \cap B) = (\sim A) \cup (\sim B)$ follows from the relation $\sim(p \wedge q) = (\sim p) \vee (\sim q)$.

EXERCISES 5.6

1. Show that if $A \subset B$ and $B \subset A$, then $A = B$.

2. Show that if $A \subset B$ and if $B \subset C$, then $A \subset C$, by using defining statements.

3. Distinguish informally, but clearly, the difference between being an element of a set and being a subset of a set.

4. Distinguish between the *empty set* and *nothing*.

5. Use the indirect method outlined in Section 5.4 to show again that the empty set is a subset of any set.

6. Let $A = \{x: -1 \le x \le 1\}$, $B = \{x: 0 \le x \le 10\}$, $C = \{0, 1, 2, 3, 4, 5\}$, $D = \{0, 1\}$.
   (a) Determine if there are subset relations between any of the above sets.
   (b) Form the following sets and for each draw a picture of your answer:

$A \cup B$,	$D \cup C$,	$B \cap C$,
$D \cup \emptyset$,	$A \cup (B \cap C)$,	$C \cup (A \cup D)$,
$(A \cap B) \cap C$,	$C \cap (A \cap B)$,	$B \cap (C \cup A)$,
$B \cup (C \cap A)$.		

   (c) Form $A \cup D$, $D \cup C$, $B \cup C$, $D \cup B$. From the inductive evidence, what would you guess about the set $R \cup S$, given that

$R \subset S$? Try to prove it. (*Hint*: If $R \subset S$, what does that tell you about the truth statements of R and S? Look at the truth table of the appropriate compound statement under the new restricted initial conditions.)

(d) Form $A \cap D$, $D \cap C$, $B \cap C$, $D \cap B$. Similarly, what is your guess about $R \cap S$, given that $R \subset S$? Try to prove it.

7. Draw the pictures of the following sets from the geometric representation of sets in the text: $C \cap (A \cup B)$, $B \cup (A \cap C)$, $(A \cup C) \cap B$

8. Show that $\sim \emptyset = U$ and that $\sim U = \emptyset$.

9. Let p and q be the defining statements for A and B and let $A \subset B$. Then p is true whenever q is true.
   (a) What do you know about the truth of $\sim p$ whenever $\sim q$ is true?
   (b) Does this tell you anything about a subset relation between $\sim A$ and $\sim B$?

*5.7 Sets Out of Control

We have alluded earlier to the fact that if we don't put certain restrictions on our sets we can run into difficulties. The most striking of these problems takes the form of various paradoxes which arise when there are no limitations on the sets we form. Russell's paradox is the earliest, having been discovered at the turn of this century by the English mathematician and logician Bertrand Russell. It deals with the difficulty that arises when we try to form sets which contain themselves as elements.

Russell's paradox goes like this: Suppose we divide all sets into two classes. Class A will consist of all "sets" which *contain* themselves as elements, and Class B will consist of all sets which *do not contain* themselves as elements.

We will call A and B "classes" to distinguish them from the "sets" which are their elements. Notice that we are using "set" *in this section only* in an illegal, unrestricted way, so that a set may contain itself as an element.

We make some observations: Every set is in either Class A or Class B, since there is no third alternative. Also, both classes are themselves sets, so, in particular, Class B must be in one or another of the two classes. Finally, we are not talking vacuously; that is, neither of the classes is empty. The set of all chairs, say, is in Class B, since the set of chairs is not, in itself, a chair. Class A is not empty, since the set of all sets is itself a set; thus it contains itself as a member.

Question: In which class does Class B belong?

If Class B belonged to Class A, then (by definition of Class A), it would contain itself as an element. That is, Class $B \in$ Class B. But this is impossible because (by definition of Class B), no set in Class B contains itself as an element.

Well then, *Class B must belong to Class B if it doesn't belong to Class A*. But Class $B \notin$ Class B, and we are back to the previous difficulty.

Thus Class B can belong *neither* to Class A nor to Class B(!); yet as we saw, it had to be either in one or the other (!!!?).

There is no cheap way out of this paradox. It is not merely a play on words which would disappear if looked at sufficiently hard. The paradox is inherent in *unrestricted* set theory, and can be dealt with only by placing limitations on the theory itself, as we have done.

EXERCISES 5.7

1. In a small town there is but one barber (who is male). By long established custom, every man either shaves himself or he doesn't, and on Farthing Day he is shaved by the barber. Who shaves the barber on Farthing Day?

2. All adjectives can be divided into disjoint classes, *egotopical* (those which describe themselves) and *heterotopical* (those which do not describe themselves). "Long" and "wet" are heterotopical, since "long" is a short word and there is nothing particularly damp about "wet" as a word. On the other hand "short" is a short word and "black" is printed in black ink, so that "black" and "short" are egotopical. Which kind of adjective is "heterotopical?"

3. Which of the two paradoxes in Exercises 1 and 2 are real, and which (if either) is a play on words?

Chapter 6

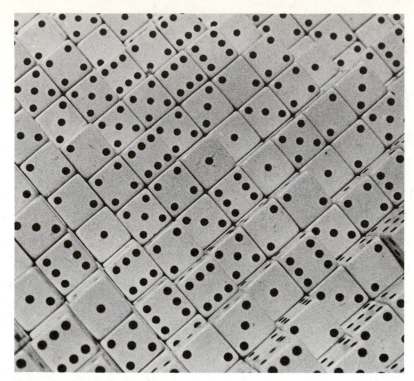

Probability

6.1 Introduction

We have all, at sometime or other, used expressions like "The probability is good that . . ." or "Probably it will . . ." Usually, what we are trying to communicate is that, under similar circumstances in the past, a certain event has occurred more often than not. Similarly, we use "The chances are poor that . . ." for something that has happened rarely, or "There is a 50–50 possibility that . . ." for a happening which occurs about as often as it does not.

Of course, it is extremely rare that the circumstances leading up to two similar events are ever *exactly* the same twice. Whenever we use expressions like those above, we are making the assumption that things are at least not too different from our previous experience. In making this assumption, we are already beginning the process of idealization.

What we are going to do in this chapter is to continue the idealization process. Our aim is to make precise many of the rather vague notions involved in "The chances are good that . . ." etc. We will do this by assigning numbers to various events according to certain rules which are based on our experience. This will then provide us with a theory of mathematical events and, as with all mathematical theories, its justification or accuracy can be judged only by comparing it with certain of the real events from which the theory was abstracted.

6.2 Events and Probabilities

The principle which governs the way in which we assign numbers to our events starts with the assumption that if certain similar conditions are repeated, then a certain event will occur some fraction (or percent) of the time.

Let's look at an example; specifically, suppose we look at the events connected with tossing a die. (A die is a cube, which has its six faces numbered 1 through 6.) If the underlying cube has been carefully constructed to balance properly, it is reasonable to suppose that, after it has been thrown, the side numbered three will appear on top (we will say that a *three comes up*) about as often as any of the others after it comes to rest. In fact, *any* one of the six sides is about as likely to come up as any of the others.

Here we are looking at six events connected with a particular experiment, namely, the tossing of a die and observing the number of the side on top after it has come to rest. *In this case* all of the six events are equally likely; that is, each of them would occur approximately one sixth of the time if we repeated the experiment a large number of times. Thus, we can reasonably assign the number $\frac{1}{6}$ to each of the events "a one is up," "a two is up," and so on.

On the other hand, suppose we were interested in whether the number appearing was less than five. In this case, we would observe, over many repetitions of tossing the die, that this event would occur about $\frac{2}{3}$ of the time, and it would be reasonable to assign the number $\frac{2}{3}$ to this event.

It is important to note what we are *not* saying here. We are not saying that, every time we repeat the experiment of tossing a die some few times, a number less than five will come up *exactly* two-thirds of the time. It is even possible that we could go through a sequence of, say, ten or more experiments of 60 tosses each without ever seeing exactly 40 times in each experiment when there would be fewer than 5 spots on top. After all, to insist that this result would occur exactly two-thirds of the time would not only require the die to "remember" what it has done so far, but also to "count" as it goes along, since it could not possibly "know" when the experiment would stop.

Observe that events we discussed in the die-tossing examples were what we might term *mathematical events*. That is, connected with each event is a statement, a *mathematical statement*, which is either true or false, with no possibility of being undecidable, or partially true. (We discussed this in Chapter 1.) Thus, the number of spots up is either less than five, or it is not. No middle ground is possible.

What is new here about these statements is the variability of their truth values; that is, even though the truth value of a describing statement must be either true or false, this value may fluctuate according to circumstances.

We can also look at events in a different way. We can consider the mathematical statements about our experiment, and then *take the truth sets of those statements for our events*. That is, we can consider that our events are *sets*! In our example a set associated in this way with "the number of spots is less than five" is {1, 2, 3, 4}, where 1 represents that a 1 is up, 2 that a 2 is up, and so on. This identification means that our problem of assigning numbers to events has been transformed into a problem of assigning numbers to *sets*, and we are at liberty to use all the machinery of set theory to help us, if we desire.

Thus, recall that we saw, in Chapter 5, that any time we deal with sets we need to encompass our problem in a universe, or universal set, in order to keep our sets under control. In this case, our universal set has a very natural interpretation; it is the set whose subsets consist of *all possible events* connected with a particular experiment. In probability theory, we call this universal set the *sample space* connected with the particular experiment, and we observe that all of our events are *subsets* of the sample space.

In our example of the die, the sample space could be represented by the set {1, 2, 3, 4, 5, 6}, where {3} represents the event that 3 spots are on top, {5} represents the event that a five is up, etc. As we have seen, the event "there are fewer than five spots up" can be represented by the set {1, 2, 3, 4}. (Note that an event has *occurred* if its defining statement is true after the experiment is over.)

The numbers we assign to the subsets of the sample space (the events) are called *probabilities*, and the whole point of probability theory is to assign probabilities to our events in a way which will be a reflection of the way the truth of their defining statements varies. Thus, if we look at the event defined by the statement "a five comes up," we assign the probability $\frac{1}{6}$ to the set {5}; or, more informally, we say, "the probability that a five comes up is $\frac{1}{6}$." Similarly, we assign $\frac{2}{3}$ to the set {1, 2, 3, 4}, or say that the probability of a number less than five coming up is $\frac{2}{3}$, since in the long run a number less than five will come up about two-thirds of the time.

This discussion leads us to the first property that we will require of the numbers which we are going to assign to our events. Since no event can occur more than 100% of the time, we stipulate that the number given to any event *can never exceed 1*. Similarly, since an event cannot happen less than zero times, we agree that our assigned numbers can never be *less than zero*. Putting these together, we see that the number given to an event, which we will call the *probability of the event*, is always between zero and one, inclusive. If we denote events by the letters *E*, *F*, *G*, etc., then their probabilities will be written $P(E)$, $P(F)$, $P(G)$, . . . , and we must have $0 \leq P(E) \leq 1$ for any event *E*. In other words the probability of a set (or event) is a number between zero and one, inclusive, expressed possibly as a percentage.

We should now take a closer look at the sample space. It would seem, theoretically at least, as if we first should construct the sample space, and then build up our events from it. In actual practice, we generally reverse the process. That is, we consider a series of related events, called the outcomes of an experiment. Thus we toss a coin (the experiment) which could have the related events (outcomes) that a *head* is on top or a *tail* is on top. Or we consider the temperature on a particular day. The experiment consists of recording the temperature to the nearest degree, and the outcomes would be integers representing the degrees on the Fahrenheit scale. We then construct our sample space by considering *all possible outcomes* of this particular experiment, as viewed from the related events which gave birth to it.

It is possible for the same experiment to have several different sample spaces, depending upon the point of view. Suppose we toss three coins, a red, a blue, and a green. If we are interested in the events describing exactly how the coins fall, we would get a sample space which could be represented by triples of the form (*H, T, T*) or (*T, H, T*), where the first triple stands for the event the red coin came down *heads*, the blue came down *tails*, and the green came down *tails*, while the second triple indicates the event that the red was *tails*, the blue *heads*, and the green *tails*. The complete sample space is the following set

$$\{(H, H, H), \quad (H, H, T), \quad (H, T, H), \quad (T, H, H),$$
$$(T, T, H), \quad (T, H, T), \quad (H, T, T), \quad (T, T, T)\}$$

On the other hand, it might be that we are interested only in counting the number of heads, without regard to the particular coins on which they appeared. In this case, the sample space would consist of only the four integers {0, 1, 2, 3}, where, say, 2 represents the event that exactly two heads appeared. Even though the experiment is the same, the related events surrounding it are different (or are at least looked at differently) and therefore give rise to *different sample spaces*.

Yet a third sample space revolving around the same experiment could come from an interest in *how often* each coin turned over while it was in the air. In this case, we might estimate that the maximum number of flips any coin would make under the circumstance is, say, 25. Just to be on the safe side, though, we would take as our sample space a set of triples, where each term in any triple is an integer from 0 to 50 inclusive. Thus, the triple (6, 0, 16) stands for the event that the *red* coin turned 6 times, the *blue* coin none, and the *green* 16.

In every case, all our sample spaces share a common property: There is no possible outcome which is not included; that is, no event *of the type considered* could take place which is *not* a subset of our sample space. Thus, a defining statement for the sample space could be "at least one of the events connected with the experiment takes place." Observe that little harm is done by taking sample spaces too large (as we have done in the third case), but nothing useful can be done with a problem if the sample space is too small.

Now, since one of the events of the sample space *must* happen, we are justified in assigning the maximum probability to the sample space (which is, after all, an event), namely, the probability of the sample space is one. Observe that, if we are taking the point of view of relative frequency, we again arrive at the assigning of the probability one to the *entire* sample space.

We shall call the events which contain but a single element of the sample space the *primitive events* of the experiment, and it is usual to begin the assigning of probabilities to the primitive events. Thus, in our first coin-tossing experiment, experience might tell us that, with fair coins, any one of the eight primitive events is as likely as any others, so we assign a probability of $\frac{1}{8}$ to each. On the other hand, experience tells us that with the second coin-tossing experiment (where we counted only the heads), "one" or "two heads" comes up about three times as often as three or none. In this case, then, we might assign probabilities $\frac{3}{8}$ to each of the events where one or two heads come up, and $\frac{1}{8}$ to each of the remaining primitive events. On the third experiment with the coins, we do not have experience to assign probabilities.

Perhaps you will have noticed the relationship between the first two examples. The primitive events of the second space can be considered to be events of the first, but not necessarily primitive. Thus the primitive event represented by {2} in the second space would be the event {(H, H, T), (H, T, H), (T, H, H)} of the first. Again we see that what is primitive and what is not depends in large measure on your point of view.

Note that the assignment of probabilities to the primitive events is not usually a mathematical problem, but is an exercise in abstracting from experience. All we have to watch out for mathematically is that all probabilities are between zero and one. Also, we have to be careful to make the assignment in such a way that the *sum of the probabilities of the primitive events* in any particular sample space is 1.

There are several characteristics to keep in mind about the primitive events in any sample space. First, notice that *no two of them can occur at the same time.* Another feature is the fact that, taking into account the point of view under consideration, each is indecomposable. That is, no primitive event can be broken down into component events which could be combined in some way. To put it differently, when we have a sample space of primitive events, there is no possible refinement consistent with the attitude taken toward that particular experiment.

For example, when we look at the second phase of our coin-tossing experiment, "exactly two heads" is no longer primitive, since it can be decomposed into the events (H, H, T), (H, T, H), and (T, H, H).

We note, in passing, that any event which has been assigned the probability one (in symbols, any event E such that $P(E) = 1$) is called a *certain event.* Any event which has probability zero (any F with $P(F) = 0$) is called the *impossible event.*[1] The event that a coin lands on its edge, for example, is an *impossible event* since all (mathematical) coins come down either heads or tails.

EXERCISES 6.2

1. Toss a die 60 times.
   (a) Record the number of 1's on top, 2's, 3's, 4's, 5's, and 6's.
   (b) Record the number of times the number of spots is less than five (use (a)).
   (c) Bring the results to class, and compare the results with those of your classmates. Also, divide the total of the results by 60 and compare with the probability.

---

[1] This terminology loses its significance if we deal with sample spaces with infinitely many elements, a difficulty which will not concern us.

2. If we say that the probability of a car having an accident at a certain corner is .01%, does this mean that for every 10,000 cars which pass the corner, there will be exactly one accident? Explain.

3. Determine which of the following are mathematical events and which are not. Since the statements occur without context, it is possible that those which appear not to be can be made mathematical in the correct setting. Where possible, find such a setting.
   (a) Four coins are tossed, and three come down heads.
   (b) John's artistic ability will increase with experience.
   (c) A coin is tossed and it comes down neither heads nor tails.
   (d) A funny thing happened on the way to class today.
   (e) His wife will be delivered of either a boy baby or a girl baby.

4. From the following physical experiments, construct or describe several sample spaces.
   (a) A coin is tossed.
   (b) Two dice are thrown.
   (c) A coin is tossed; then a die is thrown.
   (d) The following process is repeated 5 times: A telephone book is opened to a randomly chosen page, and a Ping-Pong ball numbered from 1 to 100 is drawn from a basket.
   (e) Six ball bearings are drawn from a batch, each measuring approximately 6 mm in diameter and weighing about .5 oz.

5. When possible, assign probabilities to the primitive events of the spaces in Exercise 4 (a), (b), and (c) above. Be very careful to indicate all of the assumptions you have made.

6. Toss three coins in the air and count the number of heads. Repeat this 100 times, and keep a record. Does your experience verify the claim made in the text that one or two heads will come up about thrice as often as three or none?

7. Some of the following events are certain, some are impossible, and some are neither. Which are which?
   (a) The sun rises tomorrow.
   (b) The sun sets tomorrow in the west.
   (c) A fair coin is tossed 100 times and exactly 50 heads and 50 tails appear.
   (d) A fair coin is tossed 100 times and exactly 100 tails appear.
   (e) A coin is tossed 100 times and exactly 39 heads and 51 tails appear.

6.3 New Events from Old

Just as we did with sets, we can build up new events by combining the events we have at hand in various ways. We are then able to assign probabilities to these new events in terms of the probabilities of the component events that went to make them up.

It should not surprise you that two of our combining words are *and* and *or*. That is, if we start with an event E and an event F, we can always construct two new events from them: "E and F" and "E or F." The event "E and F" is the event which can be described as "*both* of the events E and F occur," while "E or F" is defined as "*either E occurs or F occurs*," where it must be understood that we are using "or" here in its inclusive mathematical sense.

We can also use the set-theoretic point of view for defining these new events. If we consider events E and F as subsets of the sample space, then "E and F" corresponds to the set $E \cap F$, while "E or F" corresponds to $E \cup F$. Following along with this line of reasoning, we can also talk about subevents. The event G will be a *subevent* of H if G is a *subset* of H. This is the same as saying that, whenever G occurs, H must also take place, even though the converse need not hold. Keep in mind that the primitive events we talked about earlier have no proper subevents other than the empty event. This is yet another way of expressing the indecomposability of the primitive events.

A new event which can be constructed from any single event E is the event "not E." "Not E" is the event which takes place when E fails to occur, and does not happen when E does. Again, if we consider the event E as a set, then "not E" is also a set, and is, in fact, the complement of E in the sample space.

Now, how are we to go about assigning probabilities to events which are not primitive events? We have already assigned probability 1 to the event represented by the sample space, which was also *the sum of the probabilities of its component primitive subevents*. It seems reasonable, by extension, then, to do the same for any other event. Thus, the probability of E, $P(E)$, is the sum of the probabilities of the primitive events of E.

Next, suppose we have two events, E and F, which are *disjoint*; that is, events which have no primitive events in common. There are several other ways of characterizing disjoint events, all of which say the same thing. We could have said that they *cannot both happen*; that is, "E and F" is the impossible event, or we could say that $E \cap F$ is empty.

If we diagram the condition of E and F being disjoint as subsets of the sample space, we get a picture like Figure 6.1.

Figure 6.1

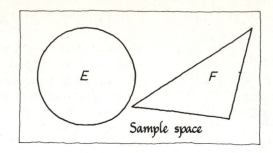

Sample space

Now a moment's reflection will convince you that the primitive events in $E \cup F$ are precisely the events in E, together with those that are in F, with no events in both. Thus, when we add up the probabilities of the primitive events in $E \cup F$, we will have precisely the sum of the probabilities of the primitive events of E plus the sum of the probabilities of the primitive events of F with nothing added in more than once. That is, for *disjoint events*,

$$P(E \cup F) = P(E \text{ or } F) = P(E) + P(F).$$

It is natural to ask whether this formula holds in the case where E and F are *not* disjoint. To explore this possibility, suppose we know that, in New York City in August, the probability that the temperature will not fall below 60° on any given day is .7 (this will be our event E), and we also know that the probability that the temperature will not exceed 90° is .8. (This is F.) Then $P(E) = .7$, $P(F) = .8$, but $P(E) + P(F) = 1.5$; this cannot be the probability of anything since it is greater than 1. The difficulty here is that "E and F" is by no means impossible; in fact, it is quite usual for the August temperature in New York City to remain between 60° and 90°.

To arrive at a general formula for $P(E \text{ or } F)$, we can see from Figure 6.2 that $E \cup F$ can be decomposed into three disjoint sets, E' (that part of E which is not in F), F' (that part of F not in E), and $E \cap F$.

Figure 6.2

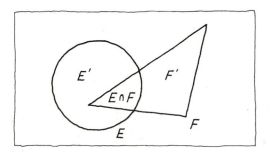

Thus the event "E or F" can be rewritten as

$$E' \quad \text{or} \quad F' \quad \text{or} \quad (E \text{ and } F)$$

where the three parts are mutually disjoint. (Here we have equated $E \cap F$ with the event "E and F".) Thus

$$P(E \text{ or } F) = P(E') + P(F') + P(E \text{ and } F)$$

But also,

$$P(E) = P(E') + P(E \text{ and } F)$$

or

$$P(E') = P(E) - P(E \text{ and } F)$$

and, similarly,

$$P(F') = P(F) - P(E \text{ and } F)$$

Substituting (since we are dealing entirely with numbers) for $P(E')$ and $P(F')$, we have

$$P(E \text{ or } F) =$$
$$P(E) - P(E \text{ and } F) + P(F) - P(E \text{ and } F) + P(E \text{ and } F)$$

or

$$P(E \text{ or } F) = P(E) + P(F) - P(E \text{ and } F)$$

This is the formula we are looking for; note that we no longer need worry about the special formula for $P(E \text{ or } F)$ for disjoint E and F. In that case, $P(E \text{ or } F) = 0$ (since $E \cup F$ is empty), so $P(E \text{ or } F)$ reduces to the previously obtained $P(E) + P(F)$.

Of special interest is what happens when we decompose the sample space into E (for any given event E) and its negative, *not E*. Since $E \cup (\sim E)$ is equal to the sample space, we have $P(E \text{ or } (\text{not } E)) = 1$. But, from the general formula (with "not E" substituted for F), we have

$$P(E \text{ or } (\text{not } E)) = P(E) + P(\text{not } E) - P(E \text{ and } (\text{not } E))$$

But

$$E \cap (\sim E) = \emptyset$$

hence $P(E \text{ and } (\text{not } E)) = 0$, which gives us

$$1 = P(E \text{ or } (\text{not } E)) = P(E) + P(\text{not } E) - 0$$

or

$$1 = P(E) + P(\text{not } E)$$

from which we see that

$$P(\text{not } E) = 1 - P(E)$$

and

$$P(E) = 1 - P(\text{not } E)$$

This last formula for $P(E)$ is useful when it may be difficult to calculate $P(E)$ directly. It allows us to find $P(E)$ by "backing" into it; first we can find $P(\text{not } E)$ and then subtract the result from 1.

We illustrate what we have been doing by an example. Suppose we throw a pair of dice, one green and one red. We will represent the result of one throw by a pair of numbers. Thus $(1, 4)$ means a 1 came up on the green and 4 came up on the red, and so on. The sample space consists of the 36 pairs given in Table 6.1.

Table 6.1

(1, 1)	(1, 2)	(1, 3)	(1, 4)	(1, 5)	(1, 6)
(2, 1)	(2, 2)	(2, 3)	(2, 4)	(2, 5)	(2, 6)
(3, 1)	(3, 2)	(3, 3)	(3, 4)	(3, 5)	(3, 6)
(4, 1)	(4, 2)	(4, 3)	(4, 4)	(4, 5)	(4, 6)
(5, 1)	(5, 2)	(5, 3)	(5, 4)	(5, 5)	(5, 6)
(6, 1)	(6, 2)	(6, 3)	(6, 4)	(6, 5)	(6, 6)

Observe the arrangement. For example, the third row represents all possible results that include a 3 on green, while the fourth column includes all possible events in which the red die came up 4.

Suppose we idealize the two dice as being completely fair. By that we mean that any one of the six numbers is as likely as any other to turn up. This in turn means that any of one of the thirty-six pairs is as probable as any of the others. Since the total must be 1, we assign to each of the pairs (our primitive events) the probability 1/36. Then the first probability that a 3 turns up on the green die is 6/36, since there are 6 pairs in the third row.

Now suppose E is the event "the sum on the faces of the two dice is 7," while F is the event "the number on the green die is even," G is "there is a three on the green die." Then $P(E) = 6/36$ (this event is composed of the primitive events on the SW–NE diagonal), $P(F) = 18/36$ (the primitive events are in rows 2, 4, and 6), and $P(G) = 6/36$.

Now $P(E \text{ or } F) = P(E) + P(F) - P(E \text{ and } F)$. "$E$ and F" contains precisely the primitive events $(6, 1)$, $(4, 3)$, and $(2, 5)$, so $P(E \text{ and } F) = 3/36$ and $P(E \text{ or } F) = 6/36 + 18/36 - 3/36 = 21/36$. This can be checked by actually counting the primitive events in Table 6.1.

Also $P(F \text{ or } G) = P(F) + P(G) - P(F \text{ and } G)$. But $P(F \text{ and } G) = 0$, since a number cannot be both 3 and even. Thus, $P(F \text{ or } G) = 18/36 + 6/36 = 24/36$. Again, check this by counting.

As a final example, to find $P(H)$, where H is "the sum on the dice is 11 or less" could be calculated directly. But it is easier to note that "not H" contains only the primitive event $(6, 6)$, so that $P(\text{not } H) = 1/36$, and therefore $P(H) = 1 - P(\text{not } H) = 1 - 1/36 = 35/36$.

EXERCISES 6.3

1. Let E be the event: A die is thrown, and the number showing is even;
   let F be the event: The die shows a 4, 5, or 6;
   let G be the event: The die shows a number divisible by 3;
   and let H be the event: The die shows a number at least equal to 3.
   (a) Describe each of the following: E or F; F and G; H or F; G and H.
   (b) Determine whether any of the events are subevents of the others. Be careful to indicate the direction of inclusion.

*2. Determine the relationship, if any, between subevents and logical implication of Chapter 5.

3. Determine which of the following pairs of events are disjoint.
   (a) A coin comes down heads; it comes down tails.
   (b) The number on the upper side of a die is even; it is divisible by 3.
   (c) It rains Monday; the sun is out Monday.
   (d) I have 10 cents in my pocket; I have 15 cents in my pocket.
   (e) The U.S. gets more gold medals at the Olympics than the USSR; the USSR gets more gold medals than the U.S. at the Olympics.
   (f) The U.S. gets as many gold medals as the USSR; the USSR gets as many gold medals as the U.S.

4. If $P(E) = .6$ and $P(F) = .9$, what do we know about $P(E$ and $F)$?

5. Show that if E, F, and G are mutually disjoint (that is, no two of them can occur at the same time), then $P(E$ or F or $G) = P(E) + P(F) + P(G)$.

6. The difficulty about calculating the probability of "E or F" can be clarified by the following exercise:
   Let $E = \{0, 1, 2, 3\}$; $F = \{2, 3, 4, 5\}$; and $G = \{4, 5\}$. Let $S(E)$ be the sum of the elements of E, etc. Thus $S(F) = 14$, and $S(G) = 9$.

(a) Calculate $S(E)$, $S(E) + S(F)$, $S(E \cup F)$, $S(E) + S(G)$, $S(E \cup G)$.

(b) Compare $S(E) + S(G)$ and $S(E \cup G)$.

(c) Compare $S(E) + S(F)$ and $S(E \cup F)$.

7. Suppose $F \subset E$. Show that $P(F)$ can be no greater than $P(E)$. How does this support your intuition about probability?

8. Referring to Table 6.1 and the accompanying text, calculate probabilities for the following events: (Do (e), (f), and (g) two ways as a check on the formulas.)

(a) The red die has a 4 showing (Event J).

(b) The sum on the two dice totals 8 (Event K).

(c) The sum on the two dice is divisible by 2. (Event L).

(d) The sum on the two dice is divisible by 3 (Event M).

(e) K or M.

(f) M or L.

(g) K or L.

9. Suppose we construct two 4-sided dice, a red one and a green one, which are both fair. (This is possible.)

(a) Construct a sample space for the experiment of throwing them both, and make a display like Table 6.1. Save it for future use.

(b) Find the probabilities of the following events. (Where calculations are involved, check your results by using the probabilities of the primitive events.)

1. A four comes up on the red die (Event E).

2. A three comes up on the green die (Event F).

3. The sum on the faces is six (Event G).

4. The sum on the faces is more than 5 (Event H).

5. E or F.

6. G or E.

7. H or F.

10. Consider the following experiment. A dictionary is opened randomly and the first letter of the first word on the left-hand page is noted.

(a) What is the obvious sample space for this problem?

(b) A little reflection will show that the 26 primitive elements of this space are *NOT* all equally likely. Why?

(c) Devise a method for assigning probabilities to the primitive elements. Execute it. Check that the sum of the probabilities of the primitive elements is one.

(d) Compare your probabilities with your classmates. Are they all the same? Who has the right answer?

6.4 Conditional Probability and the Connective "and"

As we shall see shortly, the connective *and* is not as easy to deal with as *or*. We must start with an idea which seems remote from the business of connectives, but which will eventually lead us to the calculation of $P(E$ and $F)$.

The central idea we will deal with is the notion of *conditional probability*. What we wish to do is look at the new probability of an event, given that some other event *has happened*. For example, suppose we want to know, in the two-dice problem, what is the probability that the sum is *seven*, if we know that the green has a *five* showing?

Let's look at the simplest general case first. Suppose F is a subevent of E. Then, as we know, if F has already occurred, E must have, too. Thus the new probability that E will occur, given that it is *already known* that F has occurred, is 1. We write that as $P(E|F) = 1$ [$P(E|F)$, read as *the conditional probability of E given F*]. Thus, if F is "the sum on the dice is 12," and E is "doubles were thrown," then F is a subevent of E; and if we know F occurred (the sum is 12), then we also have that doubles (double 6's) were thrown.

The basic theme running through problems in conditional probability is that, since F is given or understood to have occurred, those events which lie in the complement of F become impossible. Thus, we have an entirely new sample space, namely, F itself. In particular,

$$P(F|F) = 1 \qquad \text{or} \qquad P(F|F) = \frac{P(F)}{P(F)}$$

This gives us a clue to the solution of the more general problem in terms of the original probabilities. When we look at $P(E|F)$, we are interested in only that part of E which lies in F; that is, $E \cap F$. In terms of probability, we are concerned with that proportion of F which is taken up by $E \cap F$. Putting these together, we come up with the (mathematical) definition of conditional probability:

$$P(E|F) = \frac{P(E \text{ and } F)}{P(F)}$$

Let's check this against our two-dice problem, Table 6.1, to see if this definition will do what we want it to do. Suppose we wish to find the probability of throwing at least one three (our event E), given that the sum on the dice was seven (event F). By counting primitive events, we can see that $P(F) = 6/36$, while $P(E$ and $F) = 2/36$. Thus,

$$P(E|F) = \frac{2/36}{6/36} = 2/6 = 1/3$$

This is what we expect, seeing that 2 out of the 6 primitive possibilities satisfy the conditions of the problem.

Our formula for $P(E|F)$ has a pleasant dividend. If we multiply both sides by $P(F)$ we will have

$$P(F) \cdot P(E|F) = P(E \text{ and } F)$$

so that if we happen to be able to calculate $P(E|F)$—directly, say—we can use the information to find $P(E \text{ and } F)$.

There is, however, a somewhat general case in which we know $P(E|F)$. Suppose the events E and F are such that $P(E|F)$ is the same as $P(E)$. Again, referring to Table 6.1, take E to be "a 3 on the green die" and F to be "a 4 on the red die." Then

$$P(F) = 6/36 \text{ and } P(E \text{ and } F) = 1/36,$$

so that

$$P(E|F) = \frac{1/36}{6/36} = 1/6$$

But $P(E) = 6/36 = 1/6$, so $P(E) = P(E|F)$. Then, in this case

$$P(E \text{ and } F) = P(F) \cdot P(E) = 1/6 \times 1/6 = 1/36.$$

Again, this checks with our original methods, since $E \cap F$ contains only the single element $(3, 4)$.

Whenever we have a situation where $P(E|F) = P(E)$ or, what is equivalent (see Exercise 6.4.5), $P(F|E) = P(F)$, the events E and F are said to be *independent*. That is, E and F are independent whenever $P(E)$ is unaffected by the occurrence or nonoccurrence of F, and *vice versa*. Independent events commonly, but not exclusively, arise when we have a mechanical repetition of events with no wearing or learning process. The throwing of one die twice is an example. The die does not remember (on the second throw) what happened on the first; nor is it prescient (so that it can predict what will happen on the second throw just after the first). Therefore, any event depending only upon the first throw will be independent of any event depending only on the second.

Events which are not independent are called *dependent*. That is, if $P(E|F)$ and $P(E)$ are not the same, we say that E and F are dependent. Note that we do *not* say that E depends upon F (or *vice versa*), since *this dependence is not the same as cause and effect*.

Perhaps an example will illustrate. Suppose I wake from a nap and hear distant church bells sounding the hour of three. What is the probability that the minute hand of my watch will point between 11 and 1? If we call the first event E and the second F, then $P(F|E)$ is very close to one, assuming my watch and the church clock both keep reasonably good time. On the other hand, if I waked at a random time, the probability of the minute hand being between 11 and 1, $[P(F)]$, is $\frac{2}{12} = \frac{1}{6}$. In any case, it certainly is *not* true that $P(F|E) = P(F)$. Yet it is not reasonable to say that my watch caused the church clock to strike, nor is the opposite correct, either.

Of course, if we do have a cause-and-effect relation between events, then there is dependence; but as we have seen, the converse does not hold. That is, mathematical dependence does include cause-and-effect dependence, but it is broader. The main point to keep in mind is that two events may be dependent, in the probability sense, without there being any cause-and-effect relation between them.

EXERCISES 6.4

1. Referring to the two-dice problem, Table 6.1, and Exercise 6.3.8 calculate in two *different* ways, the probabilities of the following events, and describe them verbally.
   (a) J and K
   (b) J and M
   (c) $J|K$
   (d) $M|K$
   (e) $K|M$
   (f) $(L$ or $M)|J$

2. Referring to Exercise 6.3.9 and the table you made for it, calculate the probabilities of the following events and describe them verbally.
   (a) E and F
   (b) $E|F$
   (c) G and F
   (d) $H|E$
   (e) $(G$ or $E)|H$
   (f) $(F$ and $G)|H$

3. What did we assume about F when we calculated $P(E|F)$? Why does this reinforce our intuition about conditional probability?

4. Show that $P(E$ and $F) = P(E)P(F|E)$, also. (*Hint:* $E \cap F = F \cap E$; what does this tell you about "E and F" and "F and E"?)

5. Show that if $P(E|F) = P(E)$, then $P(F|E) = P(F)$. (*Hint:* Use Exercise 4.)

6. Show that if E and F are independent, so are E and "not F." (*Hint:* $P(\text{not } F|E) = 1 - P(F|E)$.)

7. Show that we can idealize the tossing of two coins with the same sample space as the tossing of one coin *twice*. Does this give us any clue to independent events? Extend the notion to several coins. Do your results agree with the first form of the three-coin problem in Section 6.2?

8. Determine which pairs of the four events J, K, L, and M of Exercise 6.3.8 are dependent.

9. Is it possible for two events E and F to be independent and at the same time for "E and F" to be impossible? Explain.

10. Suppose our red and green dice are loaded, so that the probability of any number coming up is proportional to the size of the number. Thus, for each die

$$P(1) = 1/21, \quad P(2) = 2/21, \quad P(3) = 3/21, \quad P(4) = 4/21,$$
$$P(5) = 5/21, \quad \text{and} \quad P(6) = 6/21.$$

(a) If a sample space for one die is $\{1, 2, 3, 4, 5, 6\}$, verify that the above probabilities, at least, may be correct for it.

(b) Use the independence of the two dice to calculate new probabilities for Table 6.1, on the basis of the above probabilities. Check that the sum of the probabilities of the primitive events is 1.

(c) Re-do Exercise 6.3.8 on the basis of the results of (b).

(d) Re-do Exercise 1 on the basis of the results of (b).

(e) Re-do Exercise 8 on the basis of (b).

11. Determine the probabilities of the primitive events of the three-coin problem if each of the coins has the property that heads is twice as probable as tails. Check your results by showing that the sum is 1. Extend the results to n coins loaded the same way.

6.5 Statistics

Closely allied with the study of probability is the branch of mathematics called statistics. In its most primitive form, statistics is the study of large batches of numbers, which provides us with a technique for making sense out of collections of data which would otherwise be incomprehensible. This can be done directly from the data itself; or, with more sophisticated techniques, intelligent guesses can be made about the large collections from relatively small samples drawn from these collections.

The raw materials of statistics are sets of numbers. The universal sets themselves are called *populations,* and subsets of the populations are often called *samples.*

When we look at any collection of numbers, it is practically automatic to ask about its average. Unfortunately, this term is so loose that it has lost any precision of meaning that it may have once had. There are at least three numbers which can be extracted from any set of data (be it a population or a sample), that might be called an average, a fact much exploited in advertising. We shall avoid the term.

The commonest number which is used to measure the middle of a set of numbers is called the *mean,* or more properly, the *arithmetic mean.* This is obtained by adding up all of the entries in the set, and then dividing by the number of elements in the data. It is probably the most familiar statistic. Another measure of the middle of a set is its *median.* The median is the "geographical center" of a set; that is, after the data has been arranged in numerical order, the median is the central element. It is the middle number, if the number of numbers is odd; it is halfway between the two middle numbers, if the number of numbers is even. Observe there are as many numbers above the median as below it. Lastly, there is the *mode,* which is that number, if there is one, which appears most often in the data. The mean and the median will always exist for any set of data, but the mode may not.

If we consider the following collection of data:

$$3, \quad 4, \quad 4, \quad 4, \quad 5, \quad 6, \quad 7, \quad 8$$

then the mean is

$$\tfrac{1}{8}(3 + 4 + 4 + 4 + 5 + 6 + 7 + 8) = 5\tfrac{1}{8}$$

the median is $\tfrac{1}{2}(4 + 5) = 4\tfrac{1}{2}$, while the mode is 4.

For a somewhat more complicated example, consider Table 6.2.

Table 6.2

Salary	Firm A	Firm B
7,000	12	11
8,000	1	1
8,700	4	9
10,000	3	1
10,700	1	1
15,000	2	0
20,000	1	1
50,000	1	1
	25	25

The table is read as follows: 12 employees of Firm A earn $7,000 per year, while only 11 employees of Firm B earn that amount; 2 employees of Firm A earn $15,000, while no one in Firm B does; and so on.

We make Table 6.3 of the three measures of centrality, for ease of comparison.

Table 6.3

	Firm A	Firm B
Mean	$10,700	$10,160
Median	8,000	8,700
Mode	7,000	7,000

Notice the wide variations of the means and the medians, even though the modes and extremes (biggest and smallest in each set) are identical.

Another important characteristic of data one often wishes to measure is the amount of dispersion it possesses. That is, is it widely scattered over a large interval, or does it tend to concentrate? A very crude measurement of dispersion is the *range* of the data (the smallest to the largest entry). This will give only the roughest of estimates, so better methods are needed.

As an illustration, consider the two sets of data

$$1, 1, 3, 4, 4, 6, 6, 7, 8, 10$$

and

$$1, 3, 4, 4, 5, 5, 6, 6, 6, 10$$

Both have the same ranges, 1 to 10, and both also have the same mean, 5. But clearly the second set is concentrated more closely about the mean than is the first.

The most useful criterion for measuring how these two sets differ is called the *standard deviation*. The standard deviation measures the way that the data is dispersed between the two extremes. If it is concentrated at the two ends, the standard deviation will be large; if it is fairly evenly spread out over the range, the standard deviation will be moderate; and if it is concentrated near the mean, the standard deviation will be small.

The easiest way to understand how this number is calculated is to see how an example is worked out by calculating the standard deviation for the first set above. We must first find the mean, which in this case is 5, and then find the difference between each entry and the mean; and then *square* it. We enter the results in Table 6.4.

Table 6.4

Differences	Squares
−4	16
−4	16
−2	4
−1	1
−1	1
1	1
1	1
2	4
3	9
5	25
0	78

(One reason we square the differences is to prevent the negative differences from canceling out the positive ones.)

We next take the mean of the squares, 7.8 in the present problem, and finally its square root, which is 2.79 to two decimal places. (Square roots in these problems are best found by use of slide rules or tables.)

We will repeat the process to find the standard deviation of the second batch of data in our example. We arrange our work as we did before, in a table (Table 6.5) for ease of calculation.

Table 6.5

Differences	Squares
−4	16
−2	4
−1	1
−1	1
0	0
0	0
1	1
1	1
1	1
5	25
0	50

The mean of the squares is 5, and the standard deviation is $\sqrt{5} = 2.24$.

Thus the two standard deviations are 2.79 and 2.24; observe that the data with the wider dispersion has the larger standard deviation. In this way, the standard deviation gives us some measure of the reliability of the mean as an indicator for judging data. If the standard deviation is small, the differences and their squares will be small, and the data is clustered about the mean, so it is a relatively good measure for the set; but if the standard deviation is large, the data is quite scattered, and the mean is a relatively poor indicator for the set of data.

The fact that the sums in the difference columns are both zero is no accident. It can be shown, with somewhat more advanced techniques, that this sum will always be zero. (It can serve as a check on your arithmetic before the next step.) This fact is one of the reasons the mean is one of the measures of centrality.

EXERCISES 6.5

1. Is there any collection of data for which the mode, the mean, and the median all exist and are equal?

2. Suppose that, in a given collection of data, all of the entries are different. What can we say about the mode?

3. Discuss any statement which starts, "Statistics prove that . . . ."

4. Check the results obtained for the means, the medians, and the modes for the data in Table 6.2.

5. In a test made by an "independent" laboratory of eight batches, each consisting of eight transistors, the following numbers of failures were found:

   Brand X: 3, 6, 4, 4, 5, 8, 7, 4.
   Brand Y: 3, 8, 6, 3, 4, 3, 6, 7.

   The laboratory decided to sell the results to both Firm X and Firm Y because they "proved" that each product was superior to the other. Ethics aside, how is this possible?

6. Find the standard deviations of the two sets of data in Exercise 5.

7. Find the mean, the median, the mode, and the standard deviation of the following data:

$$11, \quad 12, \quad 9, \quad 8, \quad 10, \quad 14, \quad 9$$

6.6 Statistics Joined to Probability

The connection between probability and statistics comes when we try to extrapolate properties of a population, based on what we know about a sample drawn from that population.

Suppose, for example, that we manufacture light bulbs and wish to determine the mean lifetime of our product. For obvious reasons, it is not possible to test the entire production to destruction in order to get the data we need. What we must do is draw a sample, find its mean, and then use that information to guess at the mean of the population.

There is a difficulty in making this guess which centers around the question "How can we be sure that our sample is truly representative of the population?" The answer is that we can't. What do we mean by being "representative of the population" when we don't really have this information about it? If, in this case, we knew what the mean of the population was, we wouldn't be running the experiment in the first place.

What we *can* do, to start, is to make sure that every number of the population has an equal chance to be selected for our sample; this is called *random sampling*. Stated mathematically, it means that the probability of selecting any one element in the population is equal to the probability of picking any other. This is not as easy to accomplish as it sounds. Even in as simple a sampling procedure as picking numbered pingpong balls from an urn (as was done in a draft-number lottery), experts have found that stirring (and other similar devices) tended to rearrange layers, rather than give a true mixing action.

Even if we are careful to select our sample as randomly as possible, there is still no guarantee that all will go well. It might happen that our sample is "pathological" (just as it is possible for a fair coin to come down heads six times in a row). The risk of this can be considerably reduced by taking a sample of sufficient size; in general, the larger the sample, the smaller the risk of pathology.

Finally, whatever the mean of our sample, we could never really expect it to be exactly the same size as the mean of our population, no matter how clever or careful we were. The best that we can hope is that the mean of our sample will be "close," in some sense, to the mean of the population.

Mathematicians have been able to give rather a precise answer to the various problems we have just formulated. This answer is an interval, which depends upon probability considerations, in addition to the mean, the standard deviation, and the sample size. If we have a sample of more than 30 items, then we can say that the probability is 95% that the mean of the population will be between

$$m - 1.96 \frac{s}{\sqrt{n}} \quad \text{and} \quad m + 1.96 \frac{s}{\sqrt{n}}$$

Here the 1.96 is a number growing out of the 95% probability; m is the sample mean; s is the standard deviation of the sample, and n is the sample size. Geometrically, we can draw the interval as in Figure 6.3 and then assert that there is a 95% probability that the population mean lies within the interval.

Figure 6.3

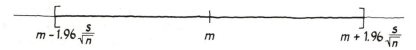

To put it differently, we are faced with two possibilities: Either,
1. The mean is in the interval; or
2. It isn't. Something unusual has happened and the mean of the population is outside of the interval.

When we say that the mean is within the interval with a *confidence level of 95%*, we mean that the probability of possibility (1) occurring is 95% while the probability of possibility (2) is only 5%.

If we wish to be more certain, that is, to increase the probability to, say, .98, we would have that the interval in which we are 98% confident the mean lies is

$$\left(m - 2.33 \frac{s}{\sqrt{n}}, \quad m + 2.33 \frac{s}{\sqrt{n}} \right)$$

and if we wish to be 99% certain, we would have to replace the coefficient 2.33 by 2.58.

There are several interesting features about the solution interval which tends to reinforce our intuition about this kind of educated guessing. We notice that, as the probability increases, so does the size of the interval. (It is as if we needed a bigger net before we could be more certain of our catch.) Also, observe that, as the sample size increases, the solution interval grows smaller though, for a substantial decrease in the interval, a relatively huge increase in the sample size is needed. We can also see that as the standard deviation increases, so does the interval; in other words, the more the dispersion present in our sample, the poorer is the mean of the sample as an indicator for the mean of the population.

However, there is another interesting point. Observe that the accuracy of the estimate in no way depends upon the size of the total population. (We do have to be careful that the sample size isn't larger than 5% of the population size.) This means that we can get quite good estimates about very large populations on the basis of samples which seem ridiculously small.

We can apply this same technique to samples of a size less than 30, but in such cases, the probability numbers, 1.96, 2.33, or 2.58 no longer remain fixed, but *vary with sample size*. Also, other facts about populations other than the mean can be estimated from samples by using techniques which are similar to the one we have just looked at. In all cases, the answers are couched in terms which use intervals and probability for estimations.

EXERCISES 6.6

1. Suppose you are sampling voters to test election preferences. Think of three ways of sampling the population, and criticize them for randomness. (That is, are there any built-in bias factors in your sample?)

2. Indicate how random sampling is a mathematical abstraction.

3. Suppose that we select, from a large production run of transistors, a sample whose mean lifetime is 150 hours, with a standard deviation of 12 hours.
   (a) Estimate the mean of the population under the following conditions, and compare all your answers.
       1. 95% confidence; sample size 100.
       2. 95% confidence; sample size 49.
       3. 95% confidence; sample size 25.
       4. 98% confidence; sample size 100.
       5. 99% confidence; sample size 100.

   (b) Verify that, if we increase the sample size to 400 (with a 95% confidence), we double the accuracy of our estimate over a1 above. (By doubling the accuracy, we mean halving the length of the estimating interval.)
   (c) How big must the sample size be, in order to double the accuracy of the estimated mean over part (b)?

REFERENCES

Huff, Durrell. *How to Lie with Statistics*. W. W. Norton & Company, Inc., New York, 1954.

Kline, Morris, ed. *Mathematics in the Modern World.* W. H. Freeman and Company, Publishers, San Francisco, 1969.

Laplace, Pierre Simon. *A Philosophical Essay on Probability.* Translated by F. W. Truscott and F. L. Emory. Dover Publications, Inc., New York, 1951.

Weaver, Warren. *Lady Luck, The Theory of Probability.* Doubleday Publishing Company, Anchor Books, Garden City, New York, 1963.

Chapter 7

Analytic Geometry and Functions

7.1 Introduction

The geometry of Euclid which we studied in Chapter 4 employed a style of proof which is often called "synthetic." In this context, *synthetic* does not have its present-day meaning of artificial, but retains its original sense from "synthesis"; that is, it describes a process of constructing a whole from its constituent parts. Specifically, in geometry, it indicates that a proof is built up from axioms and definitions.

Analytic, as used in "analytic geometry," has the converse meaning. Here it is used to denote the study of geometry by looking at—analyzing—its components. It has come to mean, moreover, that the analysis is accomplished by use of algebraic techniques. Thus, analytic geometry is the marriage of algebra and geometry; in particular, algebraic techniques are used to solve problems of geometry, and "pictures" of algebraic expressions are used to solve problems of algebra.

In this study, we will draw our axioms from a combination of axioms from geometry and algebra, and some of our definitions will take on a distinct algebraic flavor. For example, a line will be defined by means of an algebraic statement. Lines were not defined in our earlier treatment, so that as long as our lines have the appropriate properties and no conflicting ones, there will be little trouble.

The point of using algebra in geometry is that it allows us to bring to bear a tool of enormous power. This does not mean that we use it to sit back and take life easy, but—as so often happens in mathematics—we extend ourselves to take advantage of the fact that, with the new method, we can treat new and more difficult subject matter. It is true that a certain amount of purity and elegance is lost, but this is often compensated for by increased intelligibility.

7.2 Number Lines

The first step in analytic geometry is to construct the familiar *number line*, that is, to assign numbers to points, and points to numbers in a logical way, something like constructing a ruler. Our undefined objects in this case will be the real numbers themselves; and we shall assume that we are familiar with the notions of their relative size (but see Chapter 8). We will also need the concepts of the *point* and the *line* from geometry, and a way of marking off the distance between two points; all these tools are readily at hand, at least intuitively.

We can take any line whatsoever for this, and once we have chosen it, we take any two points we please and call them "zero" and "one." The distance between them determines the scale, and is called the *unit*. Custom decrees that if the line we have chosen is close to the horizontal, the point called one is to the *right* of zero; if the line is vertical, or nearly so, we usually take one *above* zero. Observe that these directional selections determine an *orientation* of the line. They are entirely arbitrary and only chosen in this way for convenience. If there is any need in a problem to reverse the practice, there is no reason that this should not be done.

Once the scale has been determined, we mark off the integers. To find the point which corresponds to *two*, we mark off one unit in the same direction as the zero-to-one interval, ordinarily to the right or upwards. To find the point corresponding to *three*, we mark off one unit to the right or upwards of two, and so on, until the positive integers all have points assigned to them, like this:

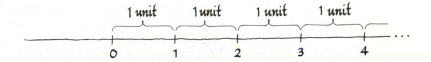

To locate the point assigned to -1, we go back to zero and mark off one unit in the direction *opposite* to the direction of one. For -2, we continue in this opposite direction one further unit, and so on, until all of the negative integers have had points assigned to them, like this:

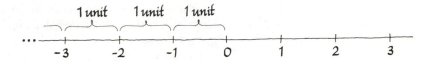

Notice that the direction of increasing size does not change; that is, if one is to the right of zero, then as we move to the right, the integers will increase, while if we go leftwards, they will decrease. This is the orientation we referred to earlier, and it will be preserved as we assign more and more numbers to our points.

We next associate points to the rationals by dividing up our integral intervals. Thus $2\frac{1}{2}$ is assigned to the point halfway between 2 and 3, while $-6\frac{2}{3}$ is assigned to the point $\frac{2}{3}$ of the way from -6 to -7, that is, ordinarily, to the left of -6 but closer to -7.

There is a neat geometric construction which allows us to divide up the intervals. Suppose, for example, we wish to locate the point $3\frac{1}{5}$. We can proceed as follows:

Figure 7.1

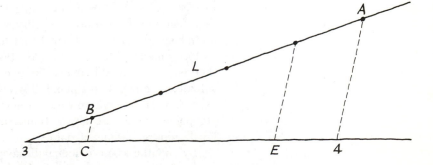

Let L be any second line through 3. Mark off 5 equal contiguous segments on L. Connect the endpoint of the last segment (A) to 4. Then through the endpoint of the first segment (B), draw a line parallel to the line $\overline{A4}$. It will intersect the segment $\overline{34}$ at some point, say C. Since triangle $3BC$ is similar to $3A4$, and since $\overline{3B}$ is one fifth of $\overline{3A}$ by construction, it follows from the geometry of similar triangles, that $\overline{3C}$ is one fifth of $\overline{34}$, and so we assign the number $3\frac{1}{5}$ to C. To find the point assigned to $3\frac{4}{5}$, we draw DE parallel to $A4$, where D is the endpoint of the fourth interval, and so on.

The method is perfectly general and can be used to find the location of the point assigned to any interval by following the basic recipe: Mark off the appropriate number of divisions (the denominator) on an auxiliary line, count off the correct number of them (the numerator), and then construct the desired similar triangle.

Finally, we need to locate the points assigned to the irrationals. Here we use the orientation. For example, suppose we wish to locate the point associated with $\sqrt{2}$. Recall that the decimal expansion for $\sqrt{2}$ is between 1.4 and 1.5, that is, between $1\frac{4}{10}$ and $1\frac{5}{10}$; hence the point belonging to $\sqrt{2}$ is between the points assigned to those fractions. Continuing, $\sqrt{2}$ lies between

$$1\frac{41}{100} \quad \text{and} \quad 1\frac{42}{100},$$

which locates it more precisely; or between

$$1\frac{414}{1000} \quad \text{and} \quad 1\frac{415}{1000},$$

or between

$$1\frac{4142}{10,000} \quad \text{and} \quad 1\frac{4143}{10,000},$$

and so on; so we can locate $\sqrt{2}$ by squeezing its position between pairs of fractions whose differences keep getting smaller.

Again, the process is quite general. We can express each irrational number as a decimal, and then locate it with ever increasing precision between pairs of rationals, one bigger and one smaller, which nevertheless approach each other as closely as we please. Observe that we are taking advantage of the fact that lines and points are abstractions, so that minute differences can be taken into account. Clearly, the difference in position between the two bounding fractions could never actually be drawn with a pencil if we were sufficiently far along in the procedure. Here is where we use the fact that the elements of our line are points, which are abstractions, rather than dots, which are the "real" representations of points.

By now the astute reader will have noticed that there are a great many points whose presence we have taken for granted. How do we know they all exist? As we pointed out in Chapter 4, on the basis of Euclid's axioms alone, we don't! We must add an axiom which gives us all the points we need; it is often called the *axiom of continuity*: On any given line there exists exactly one point for each real number. Observe that the way we assign the points to the number can be done in many ways. The selection of a scale is only choosing one manner of correspondence; with change of scale there is a change of correspondence. Nevertheless, no matter what assignment we use, the axiom assures us that there will always be enough points.

EXERCISES 7.2

1. Under what conditions would you pick zero to be "close to one"? "Remote from one"? Give practical applications in each case.

2. Give an example where it would be more convenient to select the scale with one to the left of zero. Also with one below zero.

3. It is only convenient (not necessary) to begin with points corresponding to zero and one, as the following indicate.
   (a) Show how a scale can be chosen starting with -1 and 1.
   (b) Show how a scale can be chosen starting with 0 and $\frac{3}{4}$.
   °(c) Show how a scale can be chosen starting with r and s, where r and s are *any* rational numbers.

4. Which number is smaller, -1 or $-1,000,000$? Which is (ordinarily) to the left?

5. Show how it is only necessary to work with the interval between zero and one when using the division construction to locate the rationals.

6. Actually carry out the geometric construction of the text to locate $3\frac{2}{5}$, $4\frac{5}{7}$, $-9\frac{9}{13}$.

7. What assumptions are being made about the geometry of our line when we make the division construction of the text? Why?

8. Find the numerical differences between

$$1\tfrac{4}{10} \text{ and } 1\tfrac{5}{10}, \qquad 1\tfrac{41}{100} \text{ and } 1\tfrac{42}{100},$$

$$1\tfrac{414}{1000} \text{ and } 1\tfrac{415}{1000}, \qquad 1\tfrac{4142}{10,000} \text{ and } 1\tfrac{4143}{10,000}$$

9. Locate π by using the fact that the decimal expansion for π begins 3.141593 · · · . Verify that your bounding fractions approach each other.

10. In locating points assigned to irrationals, we used the decimal expansion. Would a change of base have made any difference? Explain.

11. It is a reasonable mathematical translation of reality to define the distance between two points on a number line to be the value of the larger coordinate *minus* the value of the smaller coordinate. Find the distance between the points with coordinates
   (a) 3 and 6   (b) 6 and -12
   (c) -5 and 3   (d) -5 and -15

7.3 A Plane

The number-line concept we have just developed can easily be developed into a location device in the plane. We should expect, however, that the "address" of a point on the plane will not be as simple as a single number. On a line, once we have selected a basic reference position, the constraints are such that the location of any other point can be given by a distance and by a simple directional indicator, the plus or minus signs.

We might attempt to follow the same general scheme for locating a point on the plane. That is, we could select a base point, and then try to determine the address of a point by giving its distance from the base, and specifying a direction. This is where the difficulty lies: the direction can no longer be specified by a simple plus or minus sign. We would have to pick out a basic direction, and then give direction to our point by picking an angular measurement from the base. This method turns out to be valuable for certain problems, but for the most common applications there is a more useful method; we use number lines.

We start with two perpendicular lines, horizontal and vertical, which intersect at some convenient point. This common point is assigned to *zero* in both lines, and will be called the *origin*; the horizontal line is often called the *x axis*, the vertical line the *y axis*. A scale is selected, usually the same for both lines, with the positive numbers generally to the right on the horizontal, and upwards on the vertical line. Again, these steps are merely custom, and may be varied at the dictates of any particular problem.

To find the address of any point P on the plane, we proceed as in Figure 7.2.

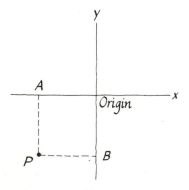

Figure 7.2

We "drop" the perpendiculars, from P, PA to the x axis and PB to the y axis. In this way *two* numbers have been assigned to P, the number assigned to the point A, where PA intersects the x axis (called the x *coordinate*) and, similarly, the point B, where PB intersects the y axis (called the y *coordinate*). If we write the x coordinate simply as x, and the y coordinate as y, then the pair of numbers x and y will be the *address* of the point P as long as we can identify which is the x and which is the y.

Again, custom decrees that we always write the pair of numbers with the x coordinate first, and the y coordinate second. We enclose the pair in parentheses to indicate that we are doing this, so that no further identification is necessary. Thus $(-3, 4)$ is the point whose x coordinate is -3 and whose y coordinate is 4. This way of writing two numbers is called an *ordered pair*, and the notation for a general ordered pair might be (x, y). Note that the point identified with $(-3, 4)$ is not the point identified with $(4, -3)$. In general, the ordered pair (x, y) is not the same as the ordered pair (y, x).

The geometric process can also be reversed. Suppose we start with a particular ordered pair, say $(-4, 3)$. We can locate the point on the plane associated with it by finding -4 on the x axis, and erecting a line perpendicular to the axis through it. Similarly, find 3 on the y axis and construct the analogous perpendicular. Where the two perpendiculars intersect is the point associated with $(-4, 3)$. This method is obviously extendable to any ordered pair (x, y).

Notice that there are many different ways of associating ordered pairs of numbers and points on the plane. Not only are there different choices of scales on the x and y axes, but we can also locate the origin at different places.

The coordinates of special points are not hard to find. Any point on the x axis must have its y coordinate zero, since any perpendicular dropped from such a point coincides with the axis itself, and intersects the y axis at its zero point. Conversely, any point whose y coordinate is zero must lie on the x axis.

Using the set notation developed in Chapter 5, we can combine these statements into a single one:

$$x \text{ axis} = \{(x, y) \mid x \text{ is any real} \quad \text{and} \quad y = 0\};$$

in words, this says that a point is on the x axis if and only if its y coordinate is zero. We can also abbreviate this by

$$x \text{ axis} = \{(x, 0) \mid x \text{ is real}\}$$

or even by saying that the x axis is given by the formula $y = 0$. All of these statements stand for the same thing.

Notice how we are utilizing the concept of the universal set here. Although we have not specifically said so, our universal set is

$$\{(x, y) \mid x \text{ is any real number and } y \text{ is any real number}\};$$

in words, the set of all ordered pairs of real numbers. Thus, when we say that the formula for the x axis is $y = 0$, we mean only that subset of the universal set whose elements have their second coordinates zero.

In a similar fashion, we see that the y axis is given by the formula $x = 0$. We now have that the origin is on *both* the x axis (so $y = 0$) and on the y axis (so $x = 0$); hence both $x = 0$ and $y = 0$; or that coordinates of the origin must be $(0, 0)$.

EXERCISES 7.3

1. On an ordinary piece of paper (not graph paper), construct a pair of number lines as indicated in the text, and geometrically locate the points with the following coordinates.
   (a) (4, 3)   (−4, 3)   (0, 1)
       (0, 0)   (0, 2)   (3, 1)
       (1/100, 6)   (6, −1,000,000)
   (b) Explain the difficulty you had in locating the last two points in (a). How did you solve this problem?

2. Go to *The New York Times, The Wall Street Journal*, or your local daily paper, and find where the idea of locating points on a plane has been utilized to make things clear. Check the horizontal and vertical scales. Are they the same? Do the number lines intersect at the zero point on both? What does each represent?

3. There is a theorem from Euclidean geometry which states that, given a line L and a point p not on L, there is only one line through p perpendicular to L. Where has this been used in the location of the point p in the plane?

4. How does an ordered pair of numbers differ from a set which contains exactly two numbers? Can you extend the notion of an ordered pair to an ordered triple? quadruple? quintuple? n-tuple?

5. What theorem of Euclidean geometry was used in locating the point assigned to the ordered pair $(−4, 3)$?

6. Carry out the details of showing that any point on the y axis must have its x coordinate zero.

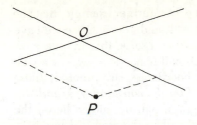

Figure 7.3

7. The axes are chosen perpendicular to each other only for convenience, and (as we shall see later) because things "look right." However, we can pick *any* two nonparallel lines as axes. Their common point is the origin, as well as zeroes for both. Pick the same scale for each. We identify points as in Figure 7.3.

From *P* draw two lines, each parallel to an axis; each will intersect a single axis at one point. This point of intersection will give the appropriate coordinate.

(a) On a plain piece of (typing) paper, draw a horizontal line (*x* axis) and a line at 45° to it (*y* axis).

(b) Mark off scales on both.

(c) Locate the points $(0, 2)$, $(3, 1)$, $(-4, 2)$, $(2, -6)$, $(-3, -2)$.

7.4 Lines and Distances

There can be no question, in analytic geometry, of writing something down, pointing to it, and then proving that it is a straight line. The reason for this is simple: the line has never been defined; it has only been described. What we *can* do, though, is to write something down (the graph of an equation in this case), and then show that it *fits the description* of a line in terms of the properties a line must have, as set down in the axioms. We will identify these graphs with lines.

To start with, suppose we look at the expression $2x + 3y = 6$. We notice that the pairs $x = 3$, $y = 0$, and $x = 0$, $y = 2$ as well as many others, make it a true statement. Using the ordered-pair notation we just developed, we could also say that $(3, 0)$ and $(0, 2)$ make it true, or *satisfy* it. Any time we have a particular pair of numbers, say x_1 and y_1, which satisfy an expression involving x and y, then we will say that (x_1, y_1) is *on the graph of that expression*. In particular, $(3, 0)$ and $(0, 2)$ are among the points on the graph of $2x + 3y = 6$.

Next, suppose we locate in the plane a large collection of points on the graph of $2x + 3y = 6$. If we have been careful in our work, the points will only *look* as if they all lie on the same straight line, partly because at this stage the notion of a line has not been defined. However, we now have a clue which will lead us to the definition.

Consider the algebraic expression

$$rx + sy = t$$

For most purposes we shall consider *r*, *s*, and *t* arbitrary but *fixed* real numbers (or constants). However, we shall need the condition that *r* and *s* are not *both* zero, although *one* of them may be. In our example above, $r = 2$, $s = 3$, and $t = 6$; notice, as a further example, that the x axis also has this form with $r = 0$, $s = 1$, and $t = 0$. Once we have fixed a particular set of numbers r, s, and t, observe that there are infinitely many number pairs (x, y) which will make $rx + sy = t$ a true statement. The set of all points whose coordinates are precisely those pairs will be called *the line determined by the constants* r, s, *and* t *in that order.* Those points (x, y) on its graph will be said to lie on the line. Notice that if we multiply r, s, and t all by the same nonzero constant K, we will have the same line, since if any pair x_1 and y_1 make $rx_1 + sy_1 = t$ a true statement, the same pair x_1 and y_1 will also make $(Kr)x_1 + (Ks)y_1 = Kt$ true, and *vice versa.* However, if r, s, and t are not proportional to, say, d, e, and f, the lines determined by the triples are not the same. Thus, $-4x - 6y = -12$ has the same graph as $2x + 3y = 6$, so that 2, 3, and 6 determine the same line as -4, -6, and -12 while the line determined by 2, 3, and 5 is different.

Now that we have objects which we have called lines, we must justify the vocabulary. That is, we must show that these objects have the desired properties: First, that two lines which are not parallel meet at just a single point; and second, that any two points determine one and only one line.[1]

Now suppose we have two lines whose equations are

$$rx + sy = t$$
$$ms + ny = p$$

We know from our work in Chapter 3 that if $rn - ms \neq 0$, there is a unique pair of numbers x and y which make both equations true. But this is exactly the statement that the point whose coordinates are this x and this y lies on both lines and (by uniqueness) only this point does this.

However, if $rn - ms = 0$, then we know there is no unique solution. Since $rn = ms$, we have, as in Chapter 3, $r/m = s/n = K$ or $r = mK$ and $s = nK$, which turns our equations into

$$(mK)x + (nK)y = t$$
$$mx + ny = p$$

or

$$(mK)x + (nK)y = t$$
$$(mK)x + (nK)y = Kp$$

so that, once again, there are solutions only if $t = Kp$. If t does equal Kp, this tells us that the two equations describe the same line, so the

---

[1]The next few paragraphs may be omitted without loss of continuity.

question of intersection does not arise. If t is not the same as Kp, then there are *no solutions*. We will show later on that this means that the two lines are parallel, as might be expected.

To show that two points determine only one line, we must turn things around. The quantities r, s, and t must be considered as the unknowns. It turns out that it will be easier to handle this problem if we rewrite $rx + sy = t$. We consider two forms of the equation, one in which $s \neq 0$ and one in which $s = 0$.

If s is *not zero*, then

$$rx + sy = t \qquad \text{Original equation}$$

$$sy = -rx + t \qquad \text{Add } -rx \text{ to both sides}$$

$$y = -\frac{r}{s}x + \frac{t}{s} \qquad \text{Divide by } s$$

$$y = mx + b \qquad \text{Rewriting in more convenient form by setting } m = -r/s \text{ and } b = t/s$$

If s is zero, then

$$rx + 0 \cdot y = t \qquad \text{Original equation}$$

$$rx = t$$

$$x = t/r \qquad \text{Dividing by } r$$

$$x = c \qquad \text{Rewriting for convenience by setting } c = t/r$$

We observe that every line equation can be put into either the form $y = mx + b$ or $x = c$, but *not both*.

Before we show in general that two points determine a single line, let's try a specific example. Let us find the line passing through $(3, 1)$ and $(-1, 2)$. Since the first x-value is 3 and the second is -1, we know that x cannot be constant. Therefore the equation of the line can have the form $y = mx + b$, where now m and b are the unknowns.

Substitution for $x = 3$ and $y = 1$ into $y = mx + b$ gives

$$1 = 3m + b$$

while $x = -1$, $y = 2$ gives

$$2 = -1 \cdot m + b$$

Subtracting, we have

$$-1 = 4m + 0 \cdot b$$

or

$$m = -\tfrac{1}{4}$$

Substituting $m = -\tfrac{1}{4}$ back into $1 = 3m + b$ gives

$$1 = -\tfrac{3}{4} + b$$

or

$$b = \tfrac{7}{4}$$

so that the equation of the line is

$$y = -\tfrac{1}{4}x + \tfrac{7}{4}$$

or, in the original $rx + sy = t$ form,

$$x + 4y = 7$$

Now suppose we are given any two fixed points (u, v) and (z, w). We must show that there is a unique line containing them both. We try the $y = mx + b$ form first, where, recall, m and b are unknowns. If we can express m and b in terms of the known values u, v, z, and w, we shall have what we are looking for.

Now, we know that

$$v = mu + b$$

and

$$w = mz + b$$

must both be true statements if there is to be a line in that form. But then we can subtract the bottom from the top, which gives

$$v - w = mu - mz$$

or

$$v - w = m(u - z) \qquad \text{Factoring the right}$$

$$\frac{v - w}{u - z} = m \qquad \text{Dividing by } u - z \text{ if } u \neq z$$

$$u = \frac{(v - w)}{(u - z)}u + b \qquad \text{Substituting for } m \text{ in the first equation}$$

$$v - \frac{v - w}{u - z}u = b \qquad \text{Adding } -\frac{(v - w)}{(u - z)}u \text{ to both sides}$$

Thus, if $u \neq z$, we have shown that there is a line through (u, v) and (z, w) whose equation has the form $y = mx + b$. We also see that the line is unique, since (recalling that m and b are the unknowns) the condition for a unique solution to the pair of equations

$$mu + b = v$$
$$mz + b = w$$

is that $u \cdot 1 - z \cdot 1 \neq 0$, which is exactly the same as saying $u \neq z$.

Next, we consider the case where $u = z$. Then, if the equations were in the $y = mx + b$ form, they would become

$$v = mu + b$$

and

$$w = mu + b$$

which gives, on subtracting, that $v - w = 0$, or $v = w$. This is impossible, since $u = z$ and $v = w$ means we don't have two points at all, just one. Therefore, if $u = z$, we must turn to the $x = c$ form of the two equations as the only one which can work. (Recall that $x = c$ is a shorthand way of indicating the set of all points whose x coordinate is c, no matter what their y coordinates are.)

The correct c for this line is, in fact, the common value of u and z; that is, the line joining (c, v) and (c, w) is $x = c$. Also, if there were another line joining our two points, we can see from the above discussion, it would have to be of the form $x = d$ for some d. But since it goes through (c, v) and (c, w), telling us that the x coordinate of every point is c, we must have $d = c$; that is, the line is unique. It is not hard to convince yourself that all the lines $x = c$ are perpendicular to the x-axis (see Exercise 7.4.9).

The concept of distance on the plane is closely tied to that of the line. Suppose, now, we consider two points on the vertical line $x = c$, say (c, v) and (c, w).

If we complete the rectangle by dropping perpendiculars to the y axis, they will intercept it at $(0, v)$ and $(0, w)$. Considered as number-line points, the distance between them is the larger coordinate minus the smaller (see Exercise 7.2.11), in our case $v - w$. But since we have a rectangle, opposite sides are equal, so $v - w$ must also be the distance between (c, v) and (c, w). That is, the distance between two points with the same x coordinate is the larger of the y coordinates minus the smaller. Similarly, the distance between two points with the same y coordinate is the larger x coordinate minus the smaller.

The numbers m and b in the $y = mx + b$ form have very special significance. The b indicates where the line crosses the y axis (see Exercise 8), and so tells us that any line in that form cannot be parallel to that axis. Now we will investigate the m.

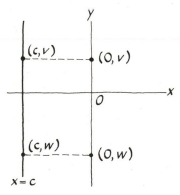

Figure 7.4

We know that if we have a line in the form $y = mx + b$, no two points on it can have the same x coordinate. Consider, now, any two points (u, v) and (z, w) on the line. Then

$$v = mu + b$$

and

$$w = mz + b$$

are both true; and once again, subtraction gives

$$v - w = mu - mz$$

or

$$\frac{v - w}{u - z} = m$$

as before. Let us draw a picture of what we have (Figure 7.5).

In our picture, we have drawn our line, our two points, and in addition, have constructed the line through (z, w) parallel to the x axis, and the line through (u, v) parallel to the y axis. All points on the line first constructed have the same y coordinate, and all on the second the same x coordinate; so the two lines must intersect at (u, w). Also, the length of the horizontal line segment is $u - z$, while the length of the vertical segment is $v - w$. From the section on trigonometry that we studied in Chapter 4, we know that

$$\tan \phi = \frac{\text{opposite}}{\text{adjacent}}$$

which, in this case, gives us

$$\tan \phi = \frac{v - w}{u - z}$$

But the previous paragraph tells us that

Figure 7.5

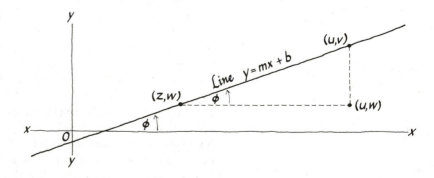

CHAPTER 7 / ANALYTIC GEOMETRY AND FUNCTIONS

$$m = \frac{v - w}{u - z}$$

so we must have that

$$m = \tan \phi$$

Since ϕ is also the angle made by our line and the x axis, we now know that *m in the equation $y = mx + b$ is the tangent of the angle made by the line and the x axis*, and this *m is called the* slope *of the line*.

In our drawing m is positive, since u is greater than z and also v is greater than w. In all cases, positive m implies that the line tilts *uphill* with increasing x like this: /. The larger m is, the steeper the rise. We can also identify lines with negative m, with lines which slope *downhill* as x increases, like this: \. In this case, if we ignore its sign, m still measures the tangent of the acute angle made by the line and the x axis. This will be justified later on.

Notice that any line of the form $x = c$ has *no slope*. This is in accord with our earlier work in trigonometry, since any such line is perpendicular to the x axis, and the angle 90° has no tangent.

We also observe that lines with the same m (slope) either coincide or are parallel to each other. This is why, earlier in this section, we ran into the possibility of either no points of intersection of two lines (they were parallel) or infinitely many such points (they coincided). This becomes clear when we recall that m depends only on the ratio $-r/s$, and not on the particular values of r and s.

To find the distance between any two points (x, v) and (z, w), we can use the Pythagorean theorem. Once again, referring to the picture (Figure 7.6) will help.

We use essentially the same picture as before, constructing the line through (x, v) and (z, w) as well as the two lines parallel to the axes. Observe that we have a right triangle with one leg having length $(x - z)$ and the other $(v - w)$. These give us that the hypotenuse (the distance we are looking for) is

$$d = \sqrt{(x - z)^2 + (v - w)^2}$$

Observe the extra leverage we get when we have the "addresses" of the points available to us. In this case, it allows us to compute the distance between two points directly and, once the scale has been selected, avoids the introduction of additional yardsticks. To put it differently, we can find distances on the plane exactly, without the necessity of approximating by use of measuring devices, once we know the scale.

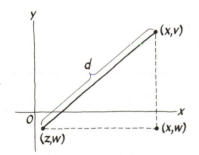

Figure 7.6

EXERCISES 7.4

1. Consider the statement $4x - y = 8$.
   (a) What are r, s, and t for this statement?
   (b) Verify that $(0, -8)$, $(2, 0)$, $(3, 4)$ all lie on the line.
   (c) Show that there are infinitely many points by verifying (inductively) that there is a point on the line for any value of x whatsoever.
   °(d) Show deductively that there are infinitely many points on the line. (*Hint*: Use (c), and proof by contradiction.)

2. Show that the points $(0, 8)$, $(2, 0)$, and $(3, 4)$ also satisfy the statement $3y - 12x = -24$. What does this fact illustrate?

3. Why is it necessary to insist that the K in the text expression $(Kr)x + (Ks)y = Kt$ be nonzero?

4. Locate the points of intersection of the following pairs of lines. Check the type of intersection expected before finding it.
   (a) $x + y = 3$ (b) $2x - 3y = 1$
       $x - y = 2$ $-x + \frac{3}{2}y = -1$
   (c) $y - 2x + 3 = 0$ (d) $x + y = 0$
       $x + 2y - 1 = 2$ $\pi x - 4y = 0$

5. What gives us the right to divide by r in deriving the form of the line equation $x = c$ in the text?

6. If you are comfortable doing a bit of algebra, show that
   (a) The expression for b in the text simplifies into $b = \dfrac{(uz - vw)}{(u - w)}$.
   (b) The same expression for b results if we substitute for m in the second equation.
   (c) Why is the result in (b) important?

7. If $m = 0$, what can you say about the line whose equation is $y = mx + b$? How is it related to the line whose equation is $x = c$?

8. Show that the b in the line equation form $y = mx + b$ is the y coordinate of the point of intersection of the line and the y axis.

9. Show that if $c \neq 0$, the line $x = c$ cannot intersect the y axis. What does this tell you about any such line? What happens if $c = 0$?

10. Uphill lines have positive slopes and downhill lines have negative slopes. What is your guess about lines which are neither uphill nor downhill? Can you verify this guess? Have you considered all possibilities?

11. Determine which of the following lines tilt uphill and which downhill, and which neither; also, where, if anywhere, they cross the axes.

(a) $y = 2x + 1$ (b) $y = 1 - 3x$
(c) $2x - y = 1$ (d) $x = 0$
(e) $y - 6x + 7 = 0$ (f) $y - 7 = 2$
(g) $x - 7y + 3 = 2x + 10y$

12. One test for parallelism of two lines was that they have the same slope. Is this the only test? Does it work for all pairs of lines?

13. Show that the distance between (u, c) and (z, c) is $u - z$ if we suppose that u is larger than z.

14. Find the distance between the following pairs of points.

(a) $(0, 2), (1, 3)$ (b) $(-1, 2), (2, 3)$
(c) $(-1, -1), (-2, -2)$

15. Is the point $(-1, 3)$ in the same line and halfway between $(2, 4)$ and $(-4, 2)$? (*Hint*: There are two ways to find out.)

7.5 Some Additional Curves

We shall start with the most familiar of all curves excepting only the straight line, the circle. We wish to find the equation of the circle whose center is at the point (h, k) with radius r. This means that we are looking for a relation between x, y, h, k, and r, so that any point whose coordinates (x, y) make the relation a true statement will be on the circle, while any other point will not.

As usual, a picture will help; see Figure 7.7.

Figure 7.7

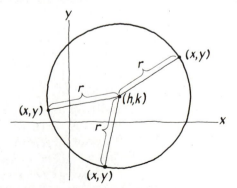

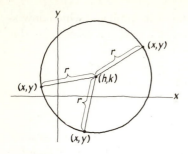

Figure 7.7

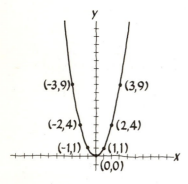

Figure 7.8

We use the definition of a circle, namely the set of all points at the distance r from (h, k). That is, if (x, y) is any point on the circle, then, using the distance formula we just derived,

$$r = \sqrt{(x - h)^2 + (y - k)^2}$$

or, after squaring both sides for convenience,

$$r^2 = (x - h)^2 + (y - k)^2$$

This turns out to be precisely the relationship we have been look- ing for. As we have seen, every point on the circle will satisfy it, while no point which is not on the circle will not because it will not satisfy the distance formula. Thus $(x - 3)^2 + (y + 2)^2 = 25$ is the circle with center at $(3, -2)$ and radius 5. Moreover, if we sketch a picture of the points satisfying the relation, we see it will even look like a circle.

Other curves will require different techniques. Often we are given a formula, and we can then try to find some of the points on the plane which satisfy it, and then draw a picture of it by connecting the points with a smooth curve. The picture is called the *graph* of the formula, and is a representation of all of the points in the plane which satisfy the formula. The number of points we plot like this will depend upon how much accuracy we need for what we are doing. This method works well enough for many purposes; but if we employ it we must be prepared for a nasty surprise once in a while. The curve may take unexpected twists between two of our plotted points.

Suppose we start with the formula $y = x^2$. This will turn out to be a parabola and, even before we start drawing its picture, our job will be easier if we make some preliminary observations. In the first place, y can never be negative (since it is the square of a number), and in fact will be zero only where x is zero.

The other point to notice is that whatever value y assumes for a particular x, it will have the same value for $-x$. This latter property is a form of symmetry. We can now set up Table 7.1.

Table 7.1

x	y	x	y
0	0	-1	1
1	1	-2	4
2	4	-3	9
3	9		

This table means that the addresses of *some* of the points on the graph of $y = x^2$ are $(0, 0)$, $(1, 1)$, $(2, 4)$, $(3, 9)$, $(-1, 1)$, $(-2, 4)$, and $(-3, 9)$. We draw in these points on our plane (Figure 7.8), and connect them with a smooth curve.

However, we may decide, for example, that since the curve seems to turn around at $(0,0)$, we need more information in its vicinity. To get this, we plot more points between $(-1, 1)$ and $(1, 1)$; for this we make Table 7.2.

Table 7.2

x	y	x	y
$\frac{3}{4}$	$\frac{9}{16}$	$-\frac{3}{4}$	$\frac{9}{16}$
$\frac{1}{2}$	$\frac{1}{4}$	$-\frac{1}{2}$	$\frac{1}{4}$
$\frac{1}{4}$	$\frac{1}{16}$	$-\frac{1}{4}$	$\frac{1}{16}$

and we magnify that portion of the picture by selecting a new scale so that the points associated with $(1, 0)$ and $(0, 1)$ are more remote from the origin:

Figure 7.9

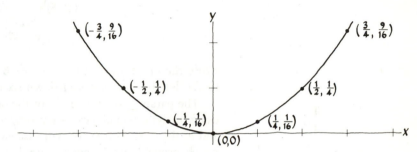

Observe that this magnification tends to confirm our larger picture, so there seems to be little point in further enlargements.

Now that we have the graph of $y = x^2$, we can use it to find other graphs. Thus $y = x^2 + 4$ will have the same shape, since every point is 4 units above the graph of $y = x^2$. Its portrait looks like Figure 7.10.

Notice that the graph of $y = x^2 - 9$ will have its turnaround point 9 units *below* the x axis, but its shape will also be the same as $y = x^2$.

The picture of $y = -x^2$ will look like a reflection of $y = x^2$; that is, it will have its pointed end up and its open end down. Otherwise it will have the same shape.

The graph of $y = (x - 3)^2$ will also look like $y = x^2$, but it will be shifted 3 units to the right, as Figure 7.11 shows,

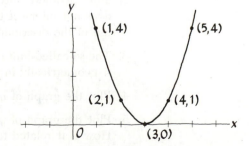

Figure 7.10

Figure 7.11

Figure 7.12

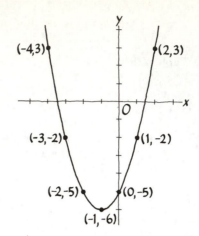

while the graph of $y = (x + 1)^2 - 6$ will be like $y = x^2$ but shifted to the left by one unit *and* down six units like Figure 7.12.

The parabola is, of course, not the only curve which can be studied using analytic techniques. Some others are included in the exercises; and in the next section we will discuss a whole family of new curves which probably could never have been examined in sufficient detail without these new techniques; yet they are of such importance that it is safe to say that modern physics could not exist without them.

EXERCISES 7.5

1. Determine the formulas for the circles with:
   (a) Radius 6 and center at (6, 5)
   (b) Radius 2 and center at $(-3, 2)$
   °(c) Through the points $(1, 0)$, $(\sqrt{2}/2, \sqrt{2}/2)$, and $(0, -1)$. Is there only one circle? How do you know? What does this illustrate in connection with Euclid's fifth postulate?

2. Show geometrically that the circle with center $(1, 4)$ and radius 2 does not intersect the x axis. Check your result algebraically by use of the discriminant.

3. The so-called *unit circle* has as its relation $x^2 + y^2 = 1$. Describe it geometrically in detail.

4. Plot the graph of $y = -x^2$, using the methods of the text.

5. Plot the graph of $y = -x^2 + 7$, using the methods of the text. How is it related to the graph of $y = -x^2$?

6. Consider the graph of $y = \sqrt{x}$.
   (a) Is y ever negative?
   (b) Is the expression meaningful for every x?
   (c) Use (a) and (b) above to decide on the general location of the graph in the plane.
   (d) Draw the graph.

°7. What does the quantity $b^2 - 4ac$ have to do with the position of the graph of $y = ax^2 + bx + c$ in relation to the x axis? (*Hint*: Go back to Chapter 3 and look at the section on quadratic equations; also look at the examples on pages 161 and 162.

8. Distinguish clearly between a curve and its portrait. It may help to reread portions of Chapter 1.

9. Sketch the graph of the following relations:
   (a) $y = (x + 2)^2$ (b) $y = (x - 7)^2$
   (c) $y = x^2 + 2x + 1$ (d) $y + 6 = (x - 2)^2$
   (e) $y + 6 = x^2 - 6x + 9$ °(f) $y = x^2 - 4x + 6$

10. Draw the pictures of the following relations. Be sure to put in enough points to satisfy yourself that you have the essentials depicted. Enlarge any interesting details.
    (a) $x = y^2$ (b) $x^2 + 4y^2 = 16$ (c) $x^2 - 4y^2 = 16$
    (d) $y = \sin x$ (Restrict your attention to x between 0° and 90°. Use Chapter 4 and a table.)

11. Consider the relation $y = 1/(x^2 + 1)$.
    (a) Is there any x for which the relation is not defined?
    (b) Is y ever bigger than one? smaller than zero? equal to zero? equal to one?
    (c) What do (a) and (b) tell you about the location of the graph of the relation?
    (d) What happens for values of x extremely remote from the origin?
    (e) Sketch the graph using all the above information.

7.6 Trigonometry Extended

We saw in Chapter 4 that trigonometry came from Euclidean geometry; one of the interesting outgrowths of analytic geometry is analytic trigonometry. Not surprisingly, it is an extension of trigonometry in much the same way that analytic geometry is an extension of geometry. In particular, it allows us to extend the notion of sin Θ, cos Θ, and tan Θ to angles Θ which measure greater than 90° or less than 0°.

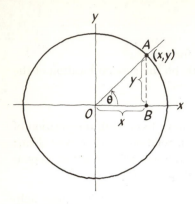

Figure 7.13

Why should anyone wish to bother? The pure mathematician might give us the classic answer, "Because it is there." But there are very practical considerations also. If we wish to describe the motion of a swinging pendulum or if we wish to find a mathematical theory of musical tone, to mention but two examples, we need to have a meaningful definition of sin Θ and cos Θ for *all* values of Θ. Notice that once we have these two functions, we have also tan Θ, since

$$\tan \Theta = \sin \Theta / \cos \Theta$$

The key to the extension of the trigonometric functions is the *unit circle*; that is, the circle with center at the origin whose radius is one. Its formula is $x^2 + y^2 = 1$. See Figure 7.13.

Now suppose we have an acute angle Θ, that is, one between 0° and 90°. We locate Θ in the plane in *standard position*, which means that one arm (called the *initial arm*) coincides with the positive x axis and the vertex is located at the origin 0. Its other arm we place so that it lies above the x axis. Called the *terminal arm*, it will cut the circle at some point A whose coordinates are (x, y).

We now complete the right triangle by dropping a perpendicular from A to the x axis where it will intersect it at B. Observe that it follows from the definition of coordinate that the length $\overline{AB} = y$ and the length $\overline{OB} = x$. Also $\overline{OA} = 1$, since it is the radius of our unit circle. Now, from the definition of sine,

$$\sin \Theta = \frac{AB}{OA} = \frac{y}{1} = y$$

and, from the definition of the cosine,

$$\cos \Theta = \frac{OB}{OA} = \frac{x}{1} = x$$

Thus, we have shown that if we place *an acute angle in standard position, the x coordinate of the point of intersection of the terminal arm and the unit circle is the angle's cosine while the y coordinate of that point is its sine.*

Observe that we have now found a way of determining sine and cosine that is free of the right triangle, and so does not depend on the angle being acute.

We now use this to extend our notions of the trigonometric functions, sin Θ and cos Θ. Consider any angle between 0° and 360°; it will be in *standard position* if one arm coincides with the positive x axis, and if its vertex is at the origin, and if we travel in a *counterclockwise* direction to reach its other (*terminal*) arm. (For our acute angles, notice that this is the same as our previous meaning.) All of the angles XOA, XOB, XOC, and XOD in Figure 7.14 illustrate angles between 0° and 360° in standard position.

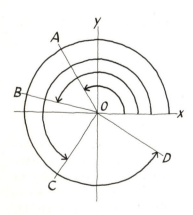

Figure 7.14

CHAPTER 7 / ANALYTIC GEOMETRY AND FUNCTIONS

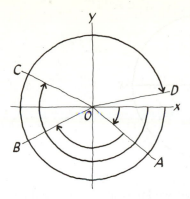

Figure 7.15

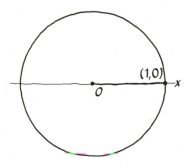

Figure 7.16

Figure 7.17

Now we can also give meaning to the idea of an angle measuring a *negative* number of degrees. Here, standard position means one arm still on the positive x axis, vertex still at (0, 0), but, as in Figure 7.15, now we proceed in a *clockwise* direction from the axis to the terminal arm. The angles XOA, XOB, XOC, and XOD are examples of angles measuring between 0° and −360°, which are in standard position.

For angles measuring larger than 360°, we just proceed from the arm on the x axis in a counterclockwise direction, making one complete circle about the origin for every time 360° divides the measure, until we reach the terminal arm. For negative angles, we proceed similarly, but in a clockwise fashion.

To define the sine and cosine of any angle Θ, we use the unit-circle property we just developed. Once the angle has been placed in standard position, we find where the terminal arm cuts the unit circle. The x coordinate of this point is defined as the cosine of Θ, while the y coordinate is defined to be the sine of Θ. Note that this gives us the same answer for an acute angle Θ as the right-triangle definition, so we are not embarrassed by having two separate definitions of sin Θ and cos Θ for the same angle Θ.

We show how to calculate some values of sin Θ and cos Θ for special measures Θ. If Θ = 0°, then the terminal arm and the positive x axis coincide (see Figure 7.16), so that $x = 1$ and $y = 0$ at the point of intersection; that is, sin 0° = 0 and cos 0° = 1.

The angle 90° (Figure 7.17) has its terminal arm coinciding with the positive y axis, making the point of intersection (0, 1), and giving us sin 90° = 1 and cos 90° = 0.

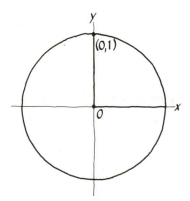

For 180°, we have that the terminal arm intersects at (−1, 0), with the result that sin 180° = 0 and cos 180° = −1 (see Figure 7.18). For −90° the point of intersection is (0, −1) producing sin (−90°) = −1 and cos (−90°) = 0.

Figure 7.18

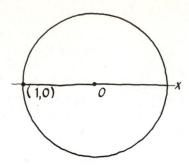

We can also take advantage of what we know about sines and co-sines of acute angles to calculate the sines and cosines of larger angles. We will use Table 7.3.

Table 7.3

Θ	$\sin \Theta$	$\cos \Theta$
30°	$\dfrac{1}{2}$	$\dfrac{\sqrt{3}}{2}$
45°	$\dfrac{\sqrt{2}}{2}$	$\dfrac{\sqrt{2}}{2}$
60°	$\dfrac{\sqrt{3}}{2}$	$\dfrac{1}{2}$

Now, suppose we want sin 120° and cos 120°. We place the angle measuring 120° in standard position, as in Figure 7.19.

Figure 7.19

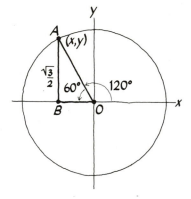

Notice that the terminal arm has swung beyond the y axis by 30°, leaving an acute angle of 60° between that arm and the x axis. Using the right triangle ABO and the fact that $\overline{AO} = 1$, we see that

$$\sin 60° = \frac{\overline{AB}}{\overline{AO}} = \frac{\overline{AB}}{1} = \overline{AB}$$

From the table $\sin 60° = \sqrt{3}/2$, so that $\overline{AB} = \sqrt{3}/2$ also $y = \overline{AB}$, so, finally, $y = \sin 120° = \dfrac{\sqrt{3}}{2}$. Similarly, $\overline{BO} = \cos 60° = \frac{1}{2}$. But $\overline{BO} = -x$, or $x = -\overline{BO}$ (since (x, y) is to the left of the x axis), so that $\cos 120° = -\frac{1}{2}$.

To find $\sin(-45°)$ and $\cos(-45°)$, once again we place the angle in standard position, as in Figure 7.20.

Figure 7.20

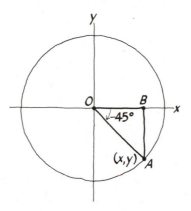

Here, we note that the acute angle made by the terminal arm and the x axis is 45°. Thus, reasoning as before, $\overline{OB} = \sqrt{2}/2$ and $\overline{BA} = \sqrt{2}/2$. This time we are to the right of the x axis and below the x axis, so that x is positive while y is negative. This gives us $\sin(-45°) = -\sqrt{2}/2$ and $\cos(-45°) = \sqrt{2}/2$.

It is easy to see that the point of intersection for the angle Θ and the angle $(\Theta + 360)°$ are the same for every Θ, since all this means is that the terminal arm has wound around the origin exactly *one more time* for the latter angle; and so the terminal arms coincide precisely. It is this fact that gives sine and cosine their most characteristic feature, their *periodicity*. That is, once we know the behavior of sine and cosine for Θ between 0° and 360°, we know about their behavior for all Θ since their values repeat every 360°.

We can use our diagram to trace the general behavior of $\sin \Theta$ as Θ increases from 0° to 360°. We start at $\Theta = 0°$, with $\sin \Theta$ (which is the vertical coordinate) at zero. Then, as Θ increases, the terminal arm swings in a counterclockwise direction. Thus, as Θ begins to increase from 0°, the $\sin \Theta$ coordinate also increases, eventually growing to the value 1 when $\Theta = 90°$. Now, as Θ continues to increase beyond

Figure 7.21

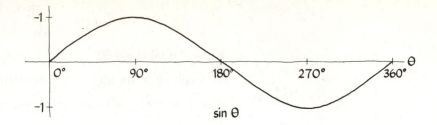

90°, the terminal arm turns beyond the positive y axis, and so (recall that our points all lie on the unit circle) the sin Θ coordinate begins to decrease, eventually becoming zero when Θ reaches 180°. As Θ grows beyond 180°, the terminal arm now is below the x axis; that is, the sin Θ coordinate becomes negative and decreases from 0 to -1 as Θ increases to 270°. Finally, as Θ goes from 270° to 360°, the sine starts to increase, growing from -1 to zero. We draw a picture of sin Θ for Θ between 0° and 360°.

Incidentally, notice how much clearer the portrait of sin Θ is than the verbal description.

EXERCISES 7.6

1. Why could sin Θ and cos Θ be defined by use of triangles only for Θ between 0° and 90°?

2. Use the material of Section 7.5 to show that the unit circle has $x^2 + y^2 = 1$ as its relation.

3. What axioms are we using in the text to insure that there *is* a point of intersection of the unit circle and the terminal arm of an angle?

4. Why is it important that the angle be placed in standard position when we deal with the sine and the cosine?

5. Follow the reasoning in the text to see why we use a circle of radius *one*, and not a circle with arbitrary radius *r*, when dealing with sine and cosine.

6. Show how the "circular" definition gives that $\sin^2\Theta + \cos^2\Theta = 1$ for any angle Θ. (*Note*: $\sin^2\Theta$ is the mathematician's way of writing $(\sin \Theta)^2$.)

7. If a clock says 3:00, give a positive and a negative angle between the hands. Are these the only two?

8. Find the sines and cosines of the following angles. Use the table in the text where necessary.
   (a) 135°   (b) −150°   (c) 315°
   (d) 225°   (e) 405°   (f) −120°

9. Draw a picture of sin Θ for Θ between:
   (a) 360° and 720°
   (b) −90° and 270°

10. Draw a picture of cos Θ, using the same techniques which were used in the text for sine Θ. Does the result look familiar? Is there any relation between sine and cosine that you know is valid for acute angles, which might account for the familiarity of your picture?

7.7 Functions

When we were taking a fresh look at algebra (Chapter 3), we discussed algebraic statements and algebraic questions. Now we will take a closer look at new mathematical objects associated with certain algebraic statements. These are called *functions*, and are considered by some mathematicians to be the most important objects in mathematics.

Actually, we have been dealing with functions throughout this chapter, and even before. The algebraic statement $y = x^2$, for example, has several functions associated with it, as we shall see. The basic idea behind the notion of the function is the concept of assigning one number to another (possibly different, possibly the same) number[1]. In our example, one function associated with $y = x^2$ assigns to any real number, x, its own square, y or x^2. Thus, 4 is assigned to $x = 2$, 9 to −3, $2\frac{1}{4}$ to $-1\frac{1}{2}$, 12 to $-2\sqrt{3}$, 1 to 1, π^3 to $\pi\sqrt{\pi}$, and so on.

Any function consists of three distinct parts, and a change in any one of them may change the entire function. The components of any function are: the *domain* of the function, that is, its allowable input, the *assignment* which specifies which number is to be associated with which number and is often, but not always, given by a formula, and the *range*, which is a list of the numbers which have had numbers assigned to them from the domain. It is the output, as it were. We can draw a picture that looks like Figure 7.22.

---

[1]This actually characterizes only one kind of function, the numerical function. However, the ideas are easily generalized, although we shall not do so here.

Figure 7.22

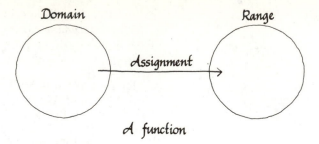

A function

Thus, in the function of our example, the assignment is given by $y = x^2$, the domain is the set of all real numbers, and the range is the set of all positive real numbers and zero.

A word of caution. Not every assignment defines a function. What distinguishes a function from its cousin, the *relation*, is that the assignment connected with a function associates a *unique* element in the range to every element in its domain, while a relation need not do so. We sometimes say that a function is *single-valued* while a relation is *multiple-valued*. Thus the assignment $y = x^2$ can be associated with a function whose domain is all real numbers, since every real number has a unique square. On the other hand, the assignment "associate every positive number with both its square roots" is not a function. Instead, it is a relation whose domain is the set of positive numbers but whose assignment is double-valued. Observe, though, that if we restrict the range of the square-root assignment to, say, the positive real numbers, *then* it becomes a function.

Whenever we have a formula where we are able to express y in terms of x alone, we say that y is *given as a function of* x, and we write it as $y = f(x)$. In this notation, y is an element of the range, x an element of the domain, and $f(x)$ stands for the assignment. The notation f *is a function* stands for the whole works, domain, assignment and range.

Often we are just given the formula associated with a particular function, without any specifications about its range and domain. In this case, we will take the domain to be the largest set of real numbers for which the formula itself is defined as a real number. This is the *natural domain* of the formula. The range in these cases will consist of exactly those numbers produced in the formula by the numbers of its natural domain. Thus, the natural domain of the formula $y = x^2 - 1$ is the set of all real numbers and its range is the set of all real numbers greater than or equal to -1. On the other hand, the natural domain of $1/(x^2 - 1)$ is the set of all reals except 1 and -1. The range is harder to figure, but more advanced techniques show it to be all reals except those between zero and -1, with zero not in the range but -1 included.

Sometimes we are given a formula and a domain which is less extensive than its natural domain. Recall, for example, that in Chapter 3 we had considered the expression

$$d = -16t^2 + 100t$$

which measured the distance d of a ball above the ground at time t. Now, as a strictly mathematical formula, d is defined for all real values of t; that is, the natural domain of d is the set of all real t. However, considered as an expression from the physical world describing a real situation, it is nonsense for any values of t which make d negative. In other words, we are interested in only those values t in the natural domain which are assigned to nonnegative values in the range. In this case, the domain of d, considered as a physical formula, is the set of t between zero and 25/4, inclusive.

In another case, the domains of our functions might be restricted to the integers, or even the positive integers. Thus, the amount of money (measured in dollars) in a bag of dimes could be given by $0.10x$, where x is some positive integer. After all, a fractional piece of a dime has no value.

Occasionally we may be given a formula involving both x and y, which we can use to solve for y as a function of x. In this case, we say that y is *implicitly given as a function of x*. Thus,

$$yx^2 + 6 = 4y - 3xy + 3x$$

can be solved for y to yield

$$y = (3x - 6)/(x^2 + 3x - 4)$$

The function f which we have found has as its domain all x except $x = 1$ and $x = -4$. (Why?) It is interesting to see what happens to our original expression for these values of x. When $x = 1$, we get $y + 6 = y + 3$, which is nonsense, while, when $x = -4$, we get $16y + 6 = 16y - 12$, which is no better.

Once again we must exercise caution. Even when we can actually solve for y, the resulting expression may not be the assignment of any function. For example,

$$x^2 + y^2 = 1$$

can be solved for y, since $y = \pm \sqrt{1 - x^2}$, but the $\pm$ sign tells us that, for each x, there are *two* y's assigned, so $x^2 + y^2 = 1$ defines y as a relation in x, but not a function of x.

No matter how we arrive at our functions, the method for graphing them proceeds as before. There is no harm, however, in using our powers of observation to get some extra information which will help us. Suppose, for example, we wish to graph

$$y = \frac{1}{x^2 + 1}$$

We first note that no element of its range is negative and, in fact, the range lies between zero (which is not included) and one (which is). (Why?) Also its domain includes all real x. These facts tell us that the graph is above the entire x axis. Also, we notice a form of symmetry we have seen before, namely, that the same y is assigned to both $-x$ and x. Further, we see that as x gets very large or very small, y gets very close to zero. Finally, before actually drawing our picture we compile Table 7.4 in order to determine the shape:

Table 7.4

x	$-x$	y
0	0	1
1	-1	$\frac{1}{2}$
2	-2	$\frac{1}{5}$
3	-3	$\frac{1}{10}$
5	-5	$\frac{1}{26}$

This gives us a picture which looks like Figure 7.23.

We make some general observations about graphing functions:

If c is not in the domain of the function, there can be no point on the graph above or below the x axis point c on the graph. To put it differently, the graph can never cross the vertical line $x = c$ when c is outside of the domain of the function.

Since no value of x is assigned to more than one y, no point on the x axis can have more than one point on the graph above it or below it.

It is often useful to locate those points, if any, where the graph crosses the axes.

Figure 7.23

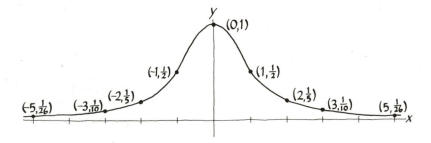

CHAPTER 7 / ANALYTIC GEOMETRY AND FUNCTIONS

Thus, when we are graphing functions, we should determine the domain first, find where the graph crosses the x axis and y axis, and find out as much about the range as possible. Then we make our table and draw in the graph.

EXERCISES 7.7

1. What number does the function associated with $y = x^2$ (as first described in the text) assign to:
   (a) $x = 3$ (b) $x = -2$ (c) $x = 4$
   (d) $x = -\frac{1}{2}$ (e) $x = \sqrt{3}$ (f) $x = -\sqrt{7}$
   (g) $x = -2\sqrt{5}$ (h) $x = -\pi$ (i) $x = a$
   (j) $x = -t$?

2. What number does a function associated with $1/(x^2 - 1)$ assign to the following numbers?
   (a) $x = 0$ (b) $x = 3$ (c) $x = -2$ (d) $x = \frac{1}{2}$
   (e) $x = -1$
   (f) What happens in (e)? What does this tell you about -1 and the domain of $\dfrac{1}{(x^2 - 1)}$?

3. What value of x has been assigned to the following numbers from the range, by a function associated with $1/(x^2 - 1)$?
   (a) $\frac{1}{3}$ (b) 1 (c) $\frac{4}{21}$ (d) $-\frac{9}{5}$ (e) $-\frac{1}{2}$ (f) 0
   (g) What do your results in (e) and (f) tell you? Do they tend to confirm any remarks made in the text concerning the range of $1/(x^2 - 1)$?

4. For each of the following formulas, find the natural domain. Where possible, find the range also. Graphing may help.
   (a) $y = x^2 + 2$ (b) $y = x^3$ (c) $y = x^3 - x$
   (d) $y = x^4 - 6$ (e) $y = \sin x$ (f) $y = \sqrt{x}$
   (g) $\sqrt{x - 1}$ (h) $y = \sqrt{-1 - x^2}$ (i) $y = \dfrac{1}{\sqrt{x}}$

   (j) $y = \dfrac{1}{x^2 - 4}$ °(k) $y = \tan x$ °(l) $y = \dfrac{1}{\sin x}$

5. In each of the following situations, specify the domain of the algebraic expression which makes the most sense.
   (a) The amount of money in a bag of quarters is $.25x$, where x is the number of quarters.

(b) A ball is thrown directly upward from a point on the ground, and the distance d above the ground at time t is given by $d = -16t^2 + 320t$.

(c) At speeds above 20 mph, the horsepower required to propel a certain ship is $h = 10{,}000s^3 + s$, where s is the speed measured in mph.

(d) The team batting average for the Red Sox in the coming season will be x.

(e) The net profit of the ABC Corp. will be x dollars this year.

(f) The next tide will vary x feet from normal.

6. Determine which of the following expressions determine y as a function of x, and which determine y as a relation in x. Find the natural domains of the functions or relations. Where possible, substitute a value for x in the expression which is *not* in the domain, to see what goes wrong.

(a) $x^2 + y^2 = 4$

(b) $yx^2 - y = 10$

(c) $y\sqrt{x^2 - 1} = 1$

(d) $y(x - 3)(x + 2) = 3x - 7$

(e) $y(x^2 - 4x + 12) = x^2 - 2x + 1$

(f) $\dfrac{y}{2x} = 3$

°(g) $y^2 + xy - 6x^2 = 0$

°(h) $y^2 + 2xy - 3x^2 + 16 = 0$

7. Consider $y = x^4$.

(a) What is its natural domain?

(b) What can you tell about its range?

(c) Where does it cross the axes?

(d) Graph it on the basis of (a), (b), (c), and a table.

8. Repeat Exercise 7 for $y = -4/(x^2 + 1)$.

9. Consider $y = x^2/(x^2 + 1)$.

(a) What is its natural domain?

(b) Is there any symmetry?

(c) What can you tell about its range, namely:

    1. Is it ever negative?

    2. Is it ever larger than 1?

    3. What happens when x is large and positive? Large and negative? (Try some samples.)

(d) Where will its graph cross the axes?

(e) Graph it.

°10. Repeat Exercise 9 for $y = x/(x^2 + 1)$. Also, in this case, the points $(1, \frac{1}{2})$ and $(-1, -\frac{1}{2})$ are interesting. Graph quite a few sample points close to each.

11. Consider $y = 1/(x - 2)$.
    (a) What is its natural domain?
    (b) What is special about the line $x = 2$, namely:
        1. Does the graph cross it?
        2. Does the graph remain a fixed distance away from it, or does it get as close as we please?
        3. Show that y is positive for x greater than 2, and negative for x less than 2.
        4. Show, by taking sample points, that y gets further from the x axis as x gets closer to 2.
    (c) Does its graph cross the axes? Where?
    (d) What happens to y when x is remote from the origin?
    (e) Graph it.

°12. Repeat Exercise 11 for $y = x/(x^2 + x - 2)$. You must find the special vertical lines yourself.

13. The unit circle (or any circle) cannot be the graph of any one function. Why?

°14. Make a table for some points on the graph of the expression $y^2 - x^3 = 0$, and try to draw its graph. Is y given implicitly as a function of x? Explain how you can tell from the graph. Is x given implicitly as a function of y? Explain.

°15. What does $b^2 - 4ac$ have to do with the domain of

$$1/(ax^2 + bx + c)?$$

REFERENCES

Loomis, Elisha Scott. *The Pythagorean Proposition*. National Council of Teachers of Mathematics, Washington, D.C., 1968.

Peter, Rozsa. *Playing with Infinity: Mathematics for Everyman*. Translated by Z. P. Dienes. Simon & Schuster, Inc., New York, 1964.

Smith, D. E., and M. L. Latham, eds. *The Geometry of René Descartes*. Dover Publications, Inc., New York, 1954.

Zippin, Leo. *Uses of Infinity*. Random House, Inc., New York, 1962.

Chapter 8

Inequalities

8.1 Introduction

Up to this time most of the algebraic relations you have studied have centered around the notion of equality. You have found numbers which made certain algebraic statements *equal* to zero, or looked for quantities which made some statements equal to other statements, and so on.

Yet surely the idea of *inequality* is one which is common and which is well worth abstracting into a mathematical theory. Our lives are circumscribed by boundaries expressed as mathematical inequalities: There are only (no more than) 24 hours in a day; we have only two hands; we can't afford (have less money than is required for) a 28-room house; the stock market must trade at least so many million shares a day to be healthy; and so on. The examples are numberless, and you can supply many of your own.

In order to provide a small taste of modern mathematics and a glimpse of the ways of the modern mathematician, we shall approach the theory of inequalities in a more formal manner than in our previous studies. After indicating our axioms and undefined notions, we shall make some formal definitions, and then proceed to state and prove theorems. After each collection of theorems, we will then provide a few examples, some worked out, and some for you to do.

8.2 Fundamentals

Since inequalities are principally concerned with numbers, we shall take the real numbers as the undefined objects for our theory. From algebra we know that the real numbers can be divided into three classes: one called the *positive numbers*, one called the *negative numbers*, and one which contains only the single number *zero*. Also, the literal symbols x, y, z, etc., will stand for numbers.

Our axioms, together with comments, are listed below. Observe that, as axioms, we cannot prove them, but they are all in accord with what we have seen in past experience.

AXIOM 8.1   *Every number falls into exactly one of the three number classes: positive, negative, or zero.*

This allows us to use the fact that every number is in one or the other of the categories, but no number can be in more than one of them.

AXIOM 8.2   *The class of positive numbers is closed under addition and multiplication. The class of negative numbers is closed under addition.*

Here, from experience, we put down in usable form the verifiable fact that the sum and product of positive numbers is positive, and that the sum of negatives is negative. Observe that it would be worse than useless to postulate that the product of negatives is negative, since this is easily disproved by our mathematical experience, as we have seen in Chapter 2. It is also unnecessary to postulate that the product of negatives is positive, since this can be proved as a theorem.

If x is positive, then −x is negative, and if x is negative, −x is positive.

Recall that −x is that number which when added to x gives zero. The familiar idea contained in this axiom, when combined with the uniqueness of the inverse, tells us that there is the "same amount" of positive numbers as negatives.

These are all the special axioms we shall need for our inequality theory. However, since the theory is an extension of the algebra we studied in Chapter 3, we will also need the eleven axioms of algebra which appeared in that chapter. It would be a good idea to go back and review them.

Before we can go ahead and prove some theorems, we will also need a few definitions. Also, a few special symbols will be useful as time and space saving devices.

DEFINITION 8.1   *We will say that x is greater than y (and write in symbols $x > y$) if $x - y$ is positive.*

Observe that this means that x is positive when and only when x is greater than zero, since x greater than zero means $x - 0$ is positive; but $x - 0 = x$, so x must be positive. Conversely, x positive means $x - 0$ positive (why), which is the same as saying (according to our definition) that $x > 0$.

DEFINITION 8.2   *We will say that w is less than z whenever $w - z$ is negative.*

The symbol for the relation "w is less than z" is $w < z$. Note that a number z is negative if and only if $z < 0$. The argument proceeds in a manner similar to the one used to show that x is positive whenever $x > 0$.

Observe that these are *definitions.* We are only stating what certain words and symbols mean. Nothing has to be proved here because there is nothing to prove. No conclusions have been drawn; we have merely delineated certain specific abstractions and singled them out for special attention.

Now we are ready to prove our first theorems.

THEOREM 8.1   *If a and b are any numbers, then exactly one of the following must hold: $a > b$, $a = b$, or $a < b$. (This property is often called the trichotomy property.)*

Proof   If we consider $a - b$, it is, according to Axiom 8.1, exactly one of the following three things: *positive*, in which case (by Definition 8.1) $a > b$; *negative*, in which case $a < b$ by Definition 8.2; or *zero*, which makes $a = b$ by the ordinary definition of equality. □[1]

THEOREM 8.2   *If $a > b$, then $b < a$, and conversely.*

Proof   Since $a > b$, we know that $a - b$ is positive. But then, by Axiom 3, $-(a - b)$ is negative, which means that $b - a$ is negative, or (Definition 8.2) $b < a$. The proof of the converse is similar. □

Theorem 8.2 is useful since it gives us the (not unexpected) relation between "greater than" and "less than." It also provides us with a strong hint that any theorem which we can prove for "greater than" will also be provable for "less than" with only minor modifications of vocabulary.

THEOREM 8.3   *If $a > b$ and $b > c$ then $a > c$. If $d < e$ and $e < f$, then $d < f$.*

This theorem expresses the orientation we have seen on the number line; if a is to the left of b, and b is to the left of c, then a is to the left of c.

Proof   We shall prove only the first of the statements and leave the second as an exercise. As $a > b$, we must have that $a - b > 0$. Similarly, $b - c > 0$, so that $(a - b) + (b - c) > 0$ (since the positive numbers are closed under addition, by Axiom 8.2). But, using the rules of algebra, $(a - b) + (b - c) = a - c$, so $a - c > 0$, or, by Definition 8.1, $a > c$. □

This completes the first group of theorems which delineates various fundamental properties of inequalities. The next batch will be "practical" theorems, that is, theorems which will give us the tools to solve

---

[1]The little square block means that the proof is finished. In many geometry books the initials Q.E.D. are used for the same thing.

problems like this: $-3x + 7 > 6 - x$. That is, we are asking if there are any numbers x which will make this inequality a true statement, and if so, what are they? Problems of this type arise not only within mathematics itself, but from the real world as well.

THEOREM 8.4   *If $a > b$, and if c is any number, then $a + c > b + c$. If $d < e$, and f is any number, then $d + f < e + f$.*

Theorem 8.4 tells us we can add the same number to both sides of an inequality without destroying or changing it.

Proof   We shall prove only the first statement and will leave the second for you to do. We have, since $a > b$, that $a - b > 0$; hence, $a - b + 0 > 0$. But $c - c = 0$, so that $a - b + c - c > 0$. A rearrangement shows us that $a + c - b - c > 0$ or $(a + c) - (b + c) > 0$, which gives that $a + c > b + c$, by Definition 8.1. □

Now that we have a theorem which says that we can add the same number to both sides of an inequality, it would be nice if we could prove a similar theorem about *multiplying* both sides. Unfortunately, here we must exercise care. We know that $2 > -3$, and if we multiply both sides of the inequality by 2, we get $4 > -6$, which is still correct. But if we multiply both sides by -2, we get $-4 > 6$, which is absurd! A little inductive experimentation will indicate that multiplying both sides by a positive number *preserves* the direction of an inequality; multiplication by a negative number will *reverse* the sense of an inequality. We prove these facts in Theorems 8.5 and 8.6.

THEOREM 8.5   *If $a > b$ and if $c > 0$, then $ac > bc$. If $d < e$ and if $f > 0$, then $fd < fe$.*

Proof   As usual, we leave the last part as an exercise. If $a > b$, then $a - b > 0$, and as $c > 0$, the closure of the positive numbers under multiplication (Axiom 8.2) tells us that $(a - b)c > 0$; hence $ac - bc > 0$ or $ac > bc$. □

THEOREM 8.6   *If $a > b$ and $c < 0$, then $ac < bc$. If $d < e$ and $f < 0$, then $fd > fe$.*

Proof   For the sake of variety, we will prove the second part and leave the first as an exercise. Since $d < e$, we have $e > d$ by Theorem 8.2. Also, by Axiom 8.3, $-f > 0$; thus $-f(e - d) > 0$, or $fd - fe > 0$, or $fd > fe$. □

A few more theorems will be useful in the next section where we will solve some problems.

THEOREM 8.7   *If $ab > 0$, then a and b are either both positive or both negative.*

This is the converse of Exercise 8.2.8(a).

Proof   We will use this to illustrate a proof by contradiction. We first observe that neither a nor b is zero. (Why?) Then, if the theorem is false, there is a pair of numbers a and b with one of the two following properties: either (1) $a > 0$, $b < 0$, and $ab > 0$, or (2) $a < 0$, $b > 0$, and $ab > 0$. Consider case 1; if $a > 0$, then multiplying both sides by b (a negative number) will *reverse* the sense of the inequality; hence, $ab < b \cdot 0$. But $b \cdot 0 = 0$, so the new inequality reads $ab < 0$, which is absurd, since the hypothesis gives us that $ab > 0$. The other case is similar. Hence the theorem must be true. □

THEOREM 8.8   *If $ab > 0$, then $a/b > 0$. Conversely, if $a/b > 0$, then $ab > 0$. That is, $a/b > 0$ when a and b are both positive or both negative.*

Proof   If $ab > 0$, then (Theorem 8.7) a and b are either both positive or both negative. If a/b is negative and if $b > 0$, then $b \cdot (a/b) < 0$, since multiplying by a positive number *preserves* the inequality. But $b(a/b) = a$; thus the inequality says that $a < 0$ while $b > 0$, which is ridiculous (by the hypothesis and Theorem 8.7). If $a/b < 0$ and $b < 0$, then $b(a/b) > 0$, since multiplying by b *reverses* the inequality. But then $a > 0$ while $b < 0$, which is also impossible. Since these are the only possible cases, we have that $a/b > 0$. □

With a little thought we see that it follows from the converse of Theorem 8.7 (Exercise 8.2.8) and Theorem 8.8 that if $a > 0$, so is $1/a$, while if $b < 0$, so is $1/b$.

THEOREM 8.9   *If $ab < 0$ then either $a > 0$ and $b < 0$ or $a < 0$ and $b > 0$.*

Proof   If a and b are both positive or both negative then $ab > 0$ (by Exercise 8.2.8(a)), which is impossible. □

THEOREM 8.10   *If $ab < 0$, then $a/b < 0$, and conversely.*

Proof   Similar to the proof of Theorem 8.9 and (for the converse), Theorem 8.8 □

Note that the conclusions of all of the above theorems are in complete agreement with our experience with the real numbers. That is, the mathematical results have already been confirmed inductively. This adds weight to our belief in the validity of our axioms, without, of course, constituting a proof.

EXERCISES 8.2

1. Is there any point in drawing up axioms about the closure of addition and multiplication in the class containing only zero? Why? State all the closure properties of this class.

2. Complete the details to show that z is negative when and only when $z < 0$.

3. Is it necessary to prove that $a > b$ if $a - b$ is positive? Why? If so, can you construct a proof?

4. Prove the second part of Theorem 8.3.

5. Prove the second part of Theorem 8.4.

6. Theorem 8.4 tells us that we may add the same number to both sides of an equality. Is it necessary to formulate a separate theorem before we can subtract the same number from both sides of an inequality? Why? If so, prove it.

°7. If $a > b$ and $c > d$, we know (inductively at least) that $a + c > b + d$. Can you prove it? Formulate a similar theorem for "less than." Can you prove it? Suppose $a > b$ and $e < f$, can you say anything about the sizes of $a + e$ and $b + f$? Can you prove it?

8. (a) Prove that if a and b have the same signs, then $ab > 0$ (we say a and b have the same signs if they are both positive or both negative). Observe that this means that the square of any nonzero number is positive. Why? How does this show that $1 > 0$?

   (b) Show, as a corollary to (a), that if a and b have different signs, then $ab < 0$.

9. Prove the second part of Theorem 8.5. You may use Exercise 8.

10. Prove the first part of Theorem 8.6. Don't forget Exercise 8.

11. Deal with the second case of Theorem 8.7.

12. The proofs of Theorem 8.8 are also proofs by contradiction. Explain the contradiction.

13. Complete the details of the proof that, if $a > 0$ and $b < 0$, then $1/a > 0$ and $1/b < 0$.

14. If $a < b$, what can be said about the sizes of $1/a$ and $1/b$? (*Hint:* Establish a result inductively first.) Can you prove your result?

15. Prove Theorem 8.10.

16. Is it necessary to prove separate theorems about dividing both sides of an inequality by the same number? Why? If so can you prove them? How is this exercise related to Exercise 6?

°17. Show that if $0 < \Theta < 180°$, then $1/\sin \Theta > 1$ or $1/\sin \Theta = 1$, and if $180° < \Theta < 360°$, then $1/\sin \Theta < -1$ or $1/\sin \Theta = -1$. Find similar results for $1/\cos \Theta$. (*Note:* $1/\sin \Theta$ is called the cosecant of Θ, and $1/\cos \Theta$ the secant of Θ. They are abbreviated csc Θ and sec Θ, respectively, and are useful in physics and engineering.)

8.3 Problem-Solving with Inequalities

We start with the simplest problems, those involving only addition and subtraction of the unknowns. Although we could present a general solution to the problems $ax + b > 0$ and $ax + b < 0$, by the time we finished with the list of particular cases ($a > 0$, $a < 0$, for each one), it would be easier to just use the rules we have developed already in each problem.

Suppose we wish to find all x's and only those x's that satisfy

$$x + 3 > 2 - x$$

To do this, we can add x to both sides, giving

$$2x + 3 > 2$$

then, subtract 3 from both sides, giving

$$2x > -1$$

and finally, dividing both sides by 2, we get

$$x > -\frac{1}{2}$$

This shows that if there are any x's which satisfy the inequality, then such an x must be greater than $-\frac{1}{2}$. We must also show that every x greater than $-\frac{1}{2}$ will satisfy the original inequality. But if $x > -\frac{1}{2}$, then $2x > -1$ (multiply both sides by 2); hence $2x + 3 > 2$ (add 3 to both sides), and therefore $x + 3 > 2 - x$ (subtract x from both sides). Thus, the original inequality is satisfied given that $x > -\frac{1}{2}$. That is, $x + 3 > 2 - x$ *when* and *only when* $x > -\frac{1}{2}$.

The last process is called *reversing the argument*; something that, in theory, should be done for every problem. The "two-way street" insures that we have precisely the right x's, not too many (we have not picked up any extras on the way which do not solve the problem) and not too few (we haven't left any out). In the future we will not go through the reversal process; the steps we will use are all reversible, but you should use the procedure yourself as one check on your work.

The exact set of x's which make a given inequality true is called the *solution set* for that inequality.

As usual, a picture will help us visualize the set. We make a sketch on the number line of the set $x > -\frac{1}{2}$ as indicated by the heavy line:

The curved bracket above the number $-\frac{1}{2}$ indicates that it is a boundary point for the solution set that is *not included* in the set itself.

Often, it is useful to explore a relation where we are interested in equality as well as strict inequality. Thus, if we know a machine can produce no more than 6 table-legs in an hour, we could say that the production is *less than or equal to* 6. The symbol for "is less then or equal to" combines the "is less than" sign and the "equals" sign into a single one, and is written $\leq$. Similarly "is greater than or equal to" is written $\geq$.

These are often called the *weak* inequalities, while the ones we have been working with earlier are the *strict* (or *strong*) inequalities.

If we wish to find the set of x's for which $\sqrt{7 - 2x}$ is defined as a real number, the answer would be the solution set of the inequality $7 - 2x \geq 0$. A little thought will convince you that all of the rules we have developed as theorems for strict inequalities can be applied to weak inequalities with only insignificant changes in vocabulary. Thus, the solution set for $7 - 2x \geq 0$ can be found as follows: Adding -7 to both sides gives $-2x \geq -7$, and dividing both sides by -2 (i.e., multiplying by $-\frac{1}{2}$) reverses the inequality so that we finally get $x \leq \frac{7}{2}$.

We sketch the solution set as we did before. In this case, the set $\{x: x \leq 7/2\}$ looks like this:

The heavy line points out the set, and the square bracket denotes a boundary point is *included* in the set.

Sometimes we are interested in exploring a relation where certain quantities are between two others. We can generally write them using a single string of symbols, such as $3 < 6 - 9x \leq 43$.

Note that when we write symbols this way the connective "and" is implied. That is,

$$3 < 6 - 9x \leq 43$$

means

$$3 < 6 - 9x \quad and \quad 6 - 9x \leq 43$$

Note, too, that either the strong or weak inequalities may be used, alone or in combination; but we must be careful that all of the inequalities go in the same direction. Thus $8 > 2 + x \geq 6$ is all right, while $6 < 2 + x \geq 8$ is meaningless. To solve string inequalities, we can operate on the three sides all at the same time. For example, to find the solution set for $-3 \leq 15 - 4x < 12$, we can just add -15 to each of the three sides, getting $-18 \leq -4x < -3$; then dividing by -4, and so *reversing* both inequalities, we have $18/4 \geq x > 3/4$. We would usually write this as $3/4 < x \leq 18/4$, since we are more comfortable going from the smaller to the larger, as we read from left to right.

A sketch of the solution set will be useful:

As before, the curved bracket means that $\frac{3}{4}$ is a boundary point that is *not* in the set, while the square bracket shows us that the boundary $\frac{18}{4}$ *is*.

We finish this section by considering simultaneous inequalities. These consist of two or more inequalities which must be satisfied at the same time. As we shall see, they are indispensable in solving inequalities involving products or quotients.

The usual state of affairs with a pair of simultaneous inequalities is that either there will be an *interval* of solutions, or the solutions will be outside of some interval, or there will be *no solutions* at all. The unique solution is relatively rare, and in some cases involving only the strong inequality, impossible.

Suppose we wish to find the set of x's which satisfies both

$$\begin{cases} 3x - 7 \geq 2 \\ 2x - 15 < -3 \end{cases}$$

Solving the first, as we have done before, we get $x \geq 3$; solving the second gives us $x < 6$.

If we sketch the two solution sets, the first just above the number line, the second below it, we have a picture which looks like this:

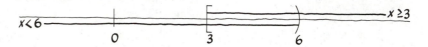

We can see that the only points which satisfy *both* inequalities are the x's in the interval $3 \leq x < 6$. In the language of sets, which we learned in Chapter 5, since any x which satisfies the pair of inequalities must be in *both* of the individual solution sets, the solution set for the pair of inequalities is the *intersection* of individual solution sets; in the previous notation,

$$\{3 \leq x < 6\} = \{x \geq 3\} \cap \{x < 6\}$$

Now, suppose we wish to solve

$$\begin{cases} 2x - 5 \geq 17 \\ 3 - x < 4 \end{cases}$$

The individual solution sets are $x \geq 11$ and $x > -1$, respectively. Sketching gives

so that the intersection of the solution sets is $x \geq 11$.

If we try to solve

$$\begin{cases} 2x - 3 \geq 1 \\ 3 - x > 6 \end{cases}$$

we get individual solution sets of $x \geq 2$ and $x < -3$. The following sketch quickly shows that the intersection of the two sets is empty, so that there is no solution to the *pair* of inequalities.

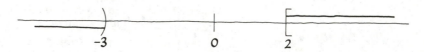

186   CHAPTER 8 / INEQUALITIES

EXERCISES 8.3

1. Find the solution set for each of the following inequalities. Sketch the solution sets.
   (a) $x - 3 > 2$
   (b) $x + 2 > 3 + 2x$
   (c) $2x + \pi < 4x - 7$
   (d) $7x + 3 - 2x > 5x - 6$

2. Find the solution set for the following weak inequalities, and sketch the solution sets.
   (a) $x + 3 \leq 4$
   (b) $x - 6 \geq 1 + 9x$
   (c) $17x - 2 \geq 21x - 19$
   (d) $-8x - 16 \leq 32 + 16x$

3. Would you use the strong or weak inequality to determine where $1/\sqrt{7 - 2x}$ is defined? Why?

4. Solve the following inequalities. Write your answers with the smallest quantities on the left. Sketch the solution sets.
   (a) $8 \leq 2x + 7 \leq 15$
   (b) $6 \geq -10x + 4 > 0$
   (c) $24 + 3x \geq 18 - 6x \geq 3x$
   (d) $15 - 2x \geq 10 - 2x > -1 - 2x$

5. Try to solve the inequality $6 \geq x + 7 \geq 8$. Sketch the solution set. Where does the trouble lie? Can you formulate a general caution from this example?

6. Solve the following *pairs* of inequalities. Sketch the solution sets in each case.
   (a) $\begin{cases} x - 2 \geq 3 \\ 2x - 7 \geq 5 \end{cases}$
   (b) $\begin{cases} 7 - x \leq 9 \\ 3x + 2 \leq 11 \end{cases}$
   (c) $\begin{cases} 7 > 13 - x \\ 2x - 4 \leq 3x - 2 \end{cases}$
   (d) $\begin{cases} -6 > 11 - x \\ x + 17 < 6 \end{cases}$

7. Show that the inequalities described as "between" in the text can be considered as a pair of simultaneous inequalities. Solve $9 + 3x \geq 6x \geq 2x - 6$.

8. The principle used to solve a pair of inequalities can be extended to solve three or more. Solve the following and sketch the solution sets.
   (a) $\begin{cases} x - 2 \geq 2 \\ 3x - 7 \geq 8 \\ x - 6 \leq 0 \end{cases}$
   (b) $\begin{cases} x - 17 > 6 \\ -6 > 11 - x \\ 10 - x < 0 \end{cases}$

9. We have used the round or curved bracket to indicate that an endpoint does not belong to a set, and a square bracket to indicate that it does. Why don't we draw in (or fail to draw in) the endpoints as needed?

8.4 Products and Quotients

We begin the study of inequalities involving products and quotients with the easiest, namely quotients of the type

$$\frac{1}{x-3} > 0$$

From an earlier section, we know that $1/(x-3)$ is positive if and only if $x-3$ is positive (that is, if and only if $x > 3$). Note that

$$\frac{1}{x-3} \geq 0$$

also has the same solution set, namely $x > 3$, since $1/(x-3)$ can never be zero.

Now, suppose we introduce an apparently insignificant change in the problem. Suppose we wish to find the solution set to

$$\frac{1}{x-3} > 1$$

We have no simple criterion, as we had before, but if we do a little algebra, perhaps something will occur to us. Subtracting 1 from both sides gives

$$\frac{1}{x-3} - 1 > 0$$

This doesn't look very encouraging. Nevertheless, we persevere with our algebra, putting everything on the left above a common denominator, and get

$$\frac{1 - (x-3)}{x-3} > 0$$

or, simplifying the numerator,

$$\frac{4-x}{x-3} > 0$$

and here, at last, is something we can use. Specifically, we can apply Theorem 8.8 of Section 8.2 which tells us that $(4-x)/(x-3)$ will be positive when and only when $4-x$ and $x-3$ are both positive or both negative. That is, we are led to several pairs of two simultaneous inequalities each:

$$\begin{cases} 4 - x > 0 \\ x - 3 > 0 \end{cases} \qquad \begin{cases} 4 - x < 0 \\ x - 3 < 0 \end{cases}$$

$$\text{or}$$
$$\begin{cases} 4 > x \\ x > 3 \end{cases} \qquad \begin{cases} 4 < x \\ x < 3 \end{cases}$$
$$\text{or} \qquad\qquad \text{or}$$
$$\begin{cases} x < 4 \\ x > 3 \end{cases} \qquad \begin{cases} x > 4 \\ x < 3 \end{cases}$$

Sketching the solution sets, we get

for the left-hand pair, which thus has a solution set $3 < x < 4$. For the right-hand pair, we have

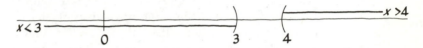

so it contributes no points to the solution set. (Recall, we are dealing with a simultaneous pair of inequalities.) Thus, the solution set for $1/(x - 3) > 1$ looks like

Now, let's try

$$\frac{2x + 3}{7 - 3x} \leq 0$$

By Theorem 8.10, we must have either $2x + 3$ is positive or zero and $7 - 3x$ is negative, or $2x + 3$ is negative or zero and $7 - 3x$ is positive. That is, we have two pairs of inequalities:

$$\begin{cases} 2x + 3 \geq 0 \\ 7 - 3x < 0 \end{cases} \qquad \begin{cases} 2x + 3 \leq 0 \\ 7 - 3x > 0 \end{cases}$$
$$\text{or} \qquad\qquad \text{or}$$
$$\begin{cases} 2x \geq -3 \\ -3x < -7 \end{cases} \qquad \begin{cases} 2x \leq -3 \\ -3x > -7 \end{cases}$$
$$\text{or} \qquad\qquad \text{or}$$
$$\begin{cases} x \geq -\frac{3}{2} \\ x > \frac{7}{3} \end{cases} \qquad \begin{cases} x \leq -\frac{3}{2} \\ x < \frac{7}{3} \end{cases}$$

A sketch of the solutions of the first pair gives

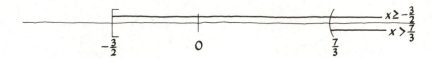

showing that $x > \frac{7}{3}$ is a solution set for it. A sketch of the solutions of the second pair looks like

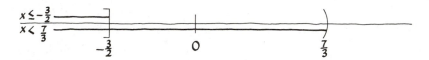

This gives us $x \leq -\frac{3}{2}$ as a solution also.

Since the original inequality $(2x - 3)/(7 - 3x) \leq 0$ is satisfied when either one pair *or* the other is, we see that a solution can come from one set *or* the other. Again, in the vocabulary and symbols of Chapter 5, the solution set for our quotient inequality is

$$\{x > \tfrac{7}{3}\} \cup \{x \leq -\tfrac{3}{2}\}$$

which looks like

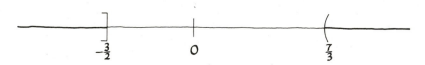

The technique for dealing with products follows along exactly the same lines, except that now we no longer have to worry about dividing by zero. For example, suppose we wish to find the solution set for

$$(x + 2)(x - 3) \leq 0$$

From Theorem 8.9 we must have that either $x + 2$ is nonnegative while $x - 3$ is nonpositive, or $x + 2$ is nonpositive while $x - 3$ is nonnegative; in symbols, this is the pair of simultaneous inequalities.

$$\begin{cases} x + 2 \geq 0 \\ x - 3 \leq 0 \end{cases} \qquad \begin{cases} x + 2 \leq 0 \\ x - 3 \geq 0 \end{cases}$$
$$\text{or} \qquad\qquad\qquad \text{or}$$
$$\begin{cases} x \geq -2 \\ x \leq 3 \end{cases} \qquad\quad \begin{cases} x \leq -2 \\ x \geq 3 \end{cases}$$

The left-hand pair has a picture which looks like

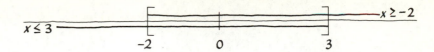

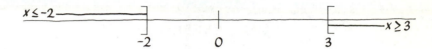

so that it contributes $-2 \leq x \leq 3$ to the solution, while the right-hand pair looks like this

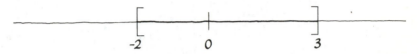

so that its part of the solution set is empty. Thus, the entire solution set is $-2 \leq x \leq 3$, which can be sketched like this:

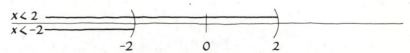

Sometimes we must factor an expression before we can begin. Thus, to solve

$$x^2 - 4 > 0$$

we first factor it into

$$(x - 2)(x + 2) > 0$$

which leads to the inequalities (Theorem 8.7)

$$\begin{cases} x - 2 > 0 \\ x + 2 > 0 \end{cases} \qquad \begin{cases} x - 2 < 0 \\ x + 2 < 0 \end{cases}$$
$$\text{or} \qquad\qquad \text{or}$$
$$\begin{cases} x > 2 \\ x > -2 \end{cases} \qquad \begin{cases} x < 2 \\ x < -2 \end{cases}$$

The left-hand pair has $x > 2$ as its solution set, as the following illustration shows:

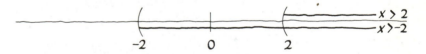

while the other pair has $x < -2$ as its solution

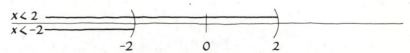

so that the solution set for $x^2 - 4 > 0$ is $\{x > 2\} \cup \{x < -2\}$, which sketches out to look like this:

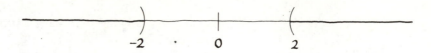

$-2 \qquad \cdot \qquad 0 \qquad 2$

EXERCISES 8.4

1. Solve the following inequalities. Sketch the solution set.

   (a) $\dfrac{1}{x + 7} > 0$ (b) $\dfrac{-3}{2x + 7} \leq 0$

   (c) $\dfrac{-1}{2x + 4} > 1$ (d) $\dfrac{x + 3}{x - 3} > 0$

   (e) $\dfrac{x + 3}{3x + 9} < 0$ (f) $\dfrac{2 - x}{2x + 71} \geq 0$

   (g) $\dfrac{1}{(x - 2)(x - 3)} < 0$

2. When we were dealing, in the text, with $(2x + 3)/(7 - 3x) \leq 0$ we used the terms nonpositive or nonnegative. Why didn't we use negative and positive respectively?

3. Solve the following inequalities and sketch the solution sets.

   (a) $(x + 7)(x - 4) > 0$ (b) $(3x - 2)(4 - 7x) \leq 0$
   (c) $(x - 3)(3 - x) < 0$ (d) $x^2 - 9 \leq 0$
   (e) $x^2 - 4x + 3 > 0$ (f) $x^2 - 6x - 7 \leq 0$
   (g) $x^2 - 5x \geq -6$

4. We can extend the techniques we have just learned to products and quotients involving more than two factors. All that is needed is to solve more than two simultaneous inequalities at the same time. Thus, to solve $(x + 1)(x + 2)(x - 3) < 0$ we note that all three factors may be negative, or any two may be positive and the remaining one negative. Thus

$$\begin{cases} x + 1 < 0 \\ x + 2 < 0 \\ x - 3 < 0 \end{cases} \quad \text{or} \quad \begin{cases} x + 1 > 0 \\ x + 2 > 0 \\ x - 3 < 0 \end{cases} \quad \text{or}$$

$$\begin{cases} x + 1 > 0 \\ x + 2 < 0 \\ x - 3 > 0 \end{cases} \quad \text{or} \quad \begin{cases} x + 1 < 0 \\ x + 2 > 0 \\ x - 3 > 0 \end{cases}$$

   (a) Complete the solution of the above inequalities.

(b) Solve the following inequalities and sketch the solution sets.

(1) $x(x - 3)(x + 4) \geq 0$ (2) $\dfrac{x + 3}{(x + 2)(x + 1)} < 0$

(3) $\dfrac{(x - 1)(x + 4)}{(x + 1)(x - 6)} \leq 0$

5. Show that $1/(x^2 + 1) > 0$ for all x. Is $1/(x_2 + 1) \leq 1$ for all x?

6. There is a geometric alternative to the solution to $(4 - x)/(x - 3) > 0$. We first find where the numerator is positive: $4 - x > 0$ or $x > 4$; then where the denominator is positive: $x > 3$. We put in a row of pluses where the numerator is positive and a row of minuses where it is negative

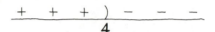

and similarly below the line for the denominator.

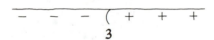

We put them together in one diagram, and get

and since the fraction is to be positive, we have, as our solution set, the region with pluses above *and* below the line. The same technique for solving $(2x + 3)/(7 - 3x) \leq 0$ would give us a picture like this.

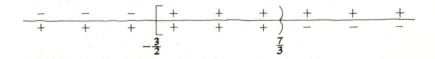

Since the fraction is to be negative, the solution set consists of regions where we have *both* a line of pluses and a line of minuses. Also note that $x = 7/3$ is *not* included in the solution set (Why?) while $x = -3/2$ is (Why?).

(a) Use this technique to solve the problems in Exercise 1.
(b) Adapt the technique to solve the problems in Exercise 3.
(c) Adapt the technique for the problems in Exercise 4.

8.5 Absolute Value

The distance *between*[1] two points a and b on the number line has been defined to be either $a - b$ (if a is the larger) or $b - a$ (if b is the larger). This definition is clumsy to use and to write, and a better way of writing it, a notation, would be a convenience, to say the least.

We could have said that the distance between a and b is

$$\begin{cases} b - a & \text{if} \quad b - a \geq 0 \\ -(b - a) & \text{if} \quad b - a < 0 \end{cases}$$

which is exactly the same as the definition of the first paragraph. Observe that by this rewritten definition, the distance between a and b is always positive or zero.

This notion of distance has been generalized to the notion of the *absolute value* of a quantity. We define the absolute value of a number A, written $|A|$ to be

$$|A| = \begin{cases} A & \text{if} \quad A \geq 0 \\ -A & \text{if} \quad A < 0 \end{cases}$$

Thus, $|3| = 3$; $|0| = 0$; $|-6| = -(-6) = 6$; and $|b - a| =$ distance between a and b without regard to which is larger.

Geometrically, the number $|A|$ represents how far the point assigned to A is from zero, measured without regard to the direction. As usual, a picture (Figure 8.1) can help.

Note that $|A|$ is positive unless $A = 0$. In particular, Axiom 8.3 tells us that whenever A is negative, then $-A = |A|$ is positive.

Figure 8.1

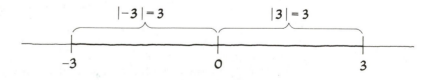

---

[1]Notice that we are *not* discussing the directed distance from a to b, which is defined to be $b - a$, allowing for the possibility of a negative quantity.

There is another way we could have defined $|A|$, namely

$$|A| = \sqrt{A^2}$$

The first thing to point out is that $A = \sqrt{A^2}$ when and only when $A \geq 0$. To see this, consider what happens when A is negative. In that case it is not possible that $A = \sqrt{A^2}$ since the left-hand side is negative while the right is positive. (Recall that when a mathematical symbol appears without a symbol preceding it, the *plus* sign is assumed to be there.) In this particular case, $\sqrt{A^2}$ is that *positive* number whose square is A^2. For example, $\sqrt{(-3)^2} = \sqrt{9} = 3$.

The square root definition gives us some valuable relationships between multiplication and absolute value.

THEOREM 8.11 $|AB| = |A|\,|B|$

Proof $\begin{aligned} |AB| &= \sqrt{(AB)^2} = \sqrt{A^2B^2} \\ &= \sqrt{A^2}\sqrt{B^2} \\ &= |A|\,|B| \quad \square \end{aligned}$

Here we have used the multiplicative property that the square root of the product is the product of the square roots.

THEOREM 8.12 $\left|\dfrac{A}{B}\right| = \dfrac{|A|}{|B|}$

Proof   Similar to Theorem 8.11. $\square$

There is no such neat relation between absolute value and addition, as the following example shows.

$$|-3 + 5| = |2| = 2$$

while

$$|-3| + |5| = 3 + 5 = 8$$

so that $|A + B|$ does not equal $|A| + |B|$ in every case. Nevertheless, we will be able to show that $|A + B|$ is not bigger then $|A| + |B|$. To do this, we need some preliminary results which are also useful on their own.

THEOREM 8.13   *Let $d > 0$. Then $|A| < d$ whenever $-d < A < d$.*
Conversely, $-d < A < d$, whenever $|A| < d$.

Figure 8.2

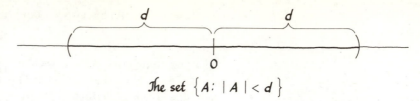

$$\text{The set } \{A: |A| < d\}$$

Proof[1]   (See Figure 8.2.) If $A \geq 0$, then, by definition, $|A| = A$, so that, given $-d < A < d$, we can conclude that $-d < |A| < d$, and in particular, $|A| < d$. If $A < 0$, then $-A = |A|$, again by definition, so that $A = -|A|$. Substituting for A in $-d < A < d$ yields $-d < -|A| < d$, and multiplying by -1 gives $d > |A| > -d$; once again we have $|A| < d$.

Now suppose $|A| < d$. If $A \geq 0$, $|A| = A$; hence $A < d$, and since $-d < 0$, $-d < A$. Putting it all together, we have $-d < 0 \leq A < d$ or (Theorem 8.3, Section 8.2) $-d < A < d$. If $A < 0$, $-A = |A|$, and substituting for $|A|$ in $|A| < d$ gives $-A < d$ or $A > -d$. But $d > 0$; hence $A < 0$ implies that $d > 0 > A$ or $d > A$. Putting it all together gives $d > A > -d$, or $-d < A < d$. $\square$

It is an easy consequence of Theorem 8.13 that a similar result holds for the weak inequality; that is, if $d > 0$, $|A| \leq d$ when and only when $-d \leq A \leq d$.

Theorem 8.13 is of more than theoretical interest. It allows us to reduce problems involving absolute values to problems involving inequalities which we already know how to solve. For example, to find the solution set for $|3x - 5| < 7$, we can rewrite it, according to Theorem 8.13 as

$$-7 < 3x - 5 < 7$$

or

$$-2 < 3x < 12$$

or

$$-\tfrac{2}{3} < x < 4$$

that is, its solution set is shown in Figure 8.3.

Figure 8.3

---

[1]This proof may be omitted

THEOREM 8.14 $|A + B| \leq |A| + |B|$

Proof[1] If $A = 0$ or $B = 0$, there is nothing to prove since if, $B = 0$ say,

$$|A + 0| = |A| = |A| + 0 = |A| + |0|$$

If both A and B are not zero, we have $A \leq |A|$, since $A = |A|$ if $A > 0$, and otherwise $A < 0$ and $|A| > 0$. Hence, by Theorem 8.13,

$$-|A| \leq A \leq |A|$$

Similarly,

$$-|B| \leq B \leq |B|$$

Adding the three sides of the inequalities (see Exercise 8.2.7), we have

$$-|A| - |B| \leq A + B \leq |A| + |B|$$

or

$$-(|A| + |B|) \leq A + B \leq |A| + |B|$$

We note that $|A| + |B|$ is positive, since it is the sum of positive numbers. Therefore we can again apply Theorem 8.13 (with $d = |A| + |B|$) and get $|A + B| \leq |A| + |B|$. □

Theorem 8.14 is called the *triangle inequality*, and confirms, mathematically, our inductive experience about distance; namely, that it is no short cut to stop off at c if we are going between a and b. To see this, the distance between a and b is $|b - a|$; the distance between a and c is $|c - a|$, while the distance between c and b is $|b - c|$. Now $|b - a|$ equals $|b - c + c - a|$ (since we can add zero in any form we like to a quantity without changing its value), and

$$|b - c + c - a| \leq |b - c| + |c - a|$$

by the triangle inequality with $A = b - c$ and $B = c - a$. Putting it all together, we get

$$|b - a| = |b - c + c - a| \leq |b - c| + |c - a|$$

or

$$|b - a| \leq |b - c| + |c - a|$$

no matter what value c has. That is, the direct route is as short as any, and shorter than some.

---

[1]This may also be omitted.

EXERCISES 8.5

1. Find the following absolute values and plot your answers on the number line.
   (a) $|5|$ (b) $|-5|$ (c) $|-32|$

   (d) $|-3| \times |2|$ (e) $|6 - 3|$ (f) $|3 - 6|$

   (g) $|-0|$ (h) $\left|\dfrac{-6}{2}\right|$ (i) $\dfrac{|-6|}{|2|}$

2. Is $-|A|$ always negative? Why?

3. Prove Theorem 8.12. What property of square roots are you using?

4. Determine what conditions are needed in order that

$$|A + B| = |A| + |B|$$

   when neither A nor B is zero. Try to prove a theorem about this.

5. (a) Show by inductive examples that if $d > 0$, then $|A| > d$ when and only when $A > d$ or $A < -d$.
   *(b) Prove (a).

6. Solve the following absolute-value inequalities. Sketch the solution
   (a) $|x - 7| < 3$ (b) $|2 - 3x| \leq 4$
   (c) $|4 - 7x| \leq 0$ (d) $|6x + 3| \geq 2$ (Use Exercise 5.)
   (e) $2 < |4 + x| < 3$

8.6 Inequalities in Two Variables[1]

There are often situations from the real world which confront us with inequalities in more than one variable. Consider, for example, a factory that is manufacturing two products, say 2-cylinder engines and 4-cylinder engines. For various reasons (availability of raw materials and components, limitation of storage facilities, transportation and shipment of finished goods, etc.), it is impractical to make more than a total of 200 units per day. If we denote by x the production of 2-cylinder engines, and by y the 4-cylinder engines, this limitation can be expressed by $x + y \leq 200$.

Now, what device is best suited to represent the solution set for this problem? Since we have a problem involving pairs of numbers,

---

[1]This section is for those who have studied some analytic geometry.

Figure 8.4

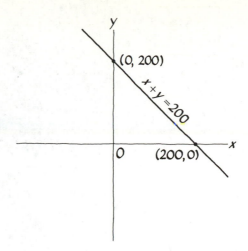

we look to the (x,y) plane, and there the solution set to our inequality, $x + y \leq 200$, will be the set of all pairs in the plane, and only those, which make $x + y \leq 200$ a true statement.

How are we to sketch the solution set? First, a little thought should convince you that the boundary of the solution set will be described when we replace the inequality by an equality. In this case, we get the line $x + y = 200$, and we sketch this on the plane in Figure 8.4.

The line divides the plane into two pieces, in our case a "northeast" part and a "southwest" part. The question now arises, which part is our solution set? The easiest way to determine this is to try a test point, since if one point in one of these regions is in the solution set they all are, while if one point of a part is not a solution, no points of that part are. Any point not on the boundary can be selected as a test point, so why not try the origin if it is available? In our case it is, since it is not a boundary point, and we see that when we substitute $x = 0$, $y = 0$ into $x + y \leq 200$ we get $0 + 0 \leq 200$ which is a true statement. Thus the origin, which is in the southwest part of the plane, is in the solution set, and, therefore, so is every other point in the southwest part, too. We indicate that by shading, as in Figure 8.5 on the next page.

One more task remains. We must decide whether the boundary is included in the solution set or not. It is quite easy to see that, in our problem, either the entire boundary is included, or none of it is. There-fore, we need use only one sample point as a test. We pick $(200,0)$ (since we know it is a boundary point) and substitute $x = 200$ and $y = 0$ into $x + y \leq 200$. We see that $200 + 0 \leq 200$ is true, and so the *entire boundary* is included in the solution set. We indicate that by drawing the boundary as a heavy solid line as in Figure 8.6.

Figure 8.5

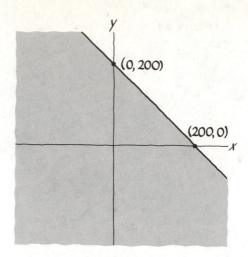

Continuing with a new problem based on the same fundamental situation, suppose the sales department insists that the number of 2-cylinder engines is at least equal to the number of 4-cylinder engines multiplied by 3/2. That is,

$$x \geq (3/2)y \qquad \text{or} \qquad 2x \geq 3y \qquad \text{or} \qquad 2x - 3y \geq 0$$

The boundary for this inequality is the line $2x - 3y = 0$, which passes through the origin and looks like Figure 8.7, where the plane is divided into northwest and southeast regions. We cannot use the origin to test which region satisfies our inequality (Why?), so we select (quite arbitrarily) the point $(-2,1)$ in the second quadrant. When we substitute $x = -2$ and $y = 1$ into $2x - 3y \geq 0$, we get $2(-2) - 3(1) \geq 0$

Figure 8.6

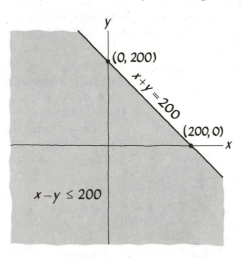

Figure 8.7

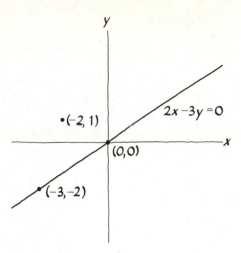

which is, of course, false. Thus the solution set is the southeastern region, and there remains only the problem of determining whether or not the boundary line lies in the set.

Here we *can* use the origin to test the boundary, and when we substitute $x = 0$ and $y = 0$ into our inequality, we see that $2(0) - (3(0) \geq 0$ is true; once again the boundary is included, and so our region looks like Figure 8.8.

If we had been dealing with the strong inequality $2x - 3y > 0$, then $2(0) - 3(0) > 0$ would have been false (Why?), and so the boundary would *not* have been in our solution set. We would have indicated this by drawing the boundary with dashes, as in Figure 8.9.

Figure 8.8

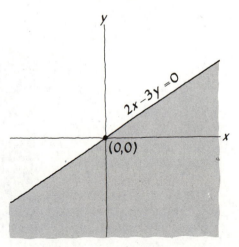

Figure 8.9

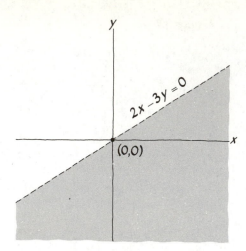

Now it is perfectly obvious from the way the problem has been developed that we are really interested in where *both* inequalities are satisfied; that is, we want to look at the *intersection* of their solution sets. To do this, we draw their pictures on the same coordinate system and get a picture which looks like Figure 8.10.

Since we want the intersection of the solution sets, we want only the region with *both* shadings, that is the darkest region which we show separately in Figure 8.11.

Furthermore, we cannot make fewer than zero engines of either type; this adds more restraints to our problem, $x \geq 0$ and $y \geq 0$. The inequality $x \geq 0$ has as its solution set the eastern half of the plane

Figure 8.10

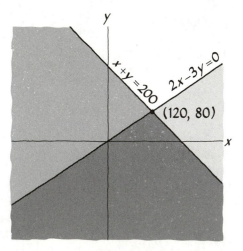

CHAPTER 8 / INEQUALITIES

Figure 8.11

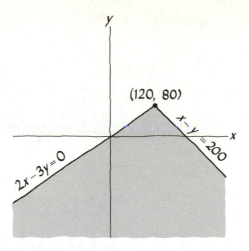

(which is to the right of the y axis), and the inequality $y \geq 0$ has as its solution set the northern region (above the x axis). Thus we are interested (for purposes of our problem) only in that portion of our previous picture which lies in the first quadrant. It looks like Figure 8.12.

Note that we have labeled all the corner points. These points are obtained by simultaneously solving the equations of the intersecting boundary lines. Thus (120,80) represents the fact that $x = 120$, $y = 80$ is the solution to the pair of equations

$$2x - 3y = 0$$
$$x + y = 200$$

Suppose, now, there is a breakdown in the plant so that production of the 4-cylinder engine is limited to a maximum of 60 units per day. This puts yet another restraint on the problem $y \leq 60$, so the intersection of all the solution sets now looks like Figure 8.13.

Note that any point of the solution set represents a possible "mix" in the production of two kinds of engines. We can now put this to

Figure 8.12

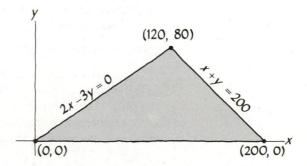

Figure 8.13

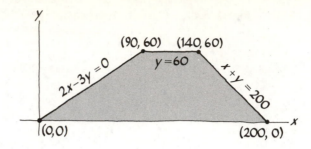

work for us, to help make certain basic decisions about the most advantageous assortment to manufacture.

For example, suppose the 2-cylinder engine sells for $600 and the 4-cylinder engine for $1000. Then the gross income per day from the factory will be $600x + 1000y$ dollars. It is an interesting fact, which can be proved by using advanced techniques, that expressions of the form $ax + by + c$ assume their largest (or maximum) values at the corner points of sets of this type. In our case, there are four to test to see which will give the maximum gross income; but commonsense and observation, in this case, also tell us that we will get the largest income when we make as many 4-cylinder engines as possible (in this example, 60). Therefore, we really need to test only (90,60) and (140,60). At (90,60) the income is

$$600(90) + 1000(60) = 114{,}000$$

while at (140,60) the income is

$$600(140) + 1000(60) = 144{,}000$$

Thus, in order to maximize gross income, the factory should produce 140 two-cylinder engines.

Now, suppose it costs $400 to make the 2-cylinder engine and $850 to make the 4-cylinder; furthermore, suppose there are fixed costs (taxes, insurance, maintenance, etc.) of $500 per day. Then the total daily cost to run the plant is $400x + 850y + 500$, so the daily profit (income less cost) will be

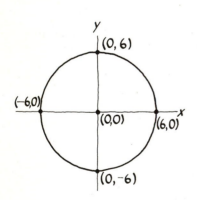

$$(600x + 1000y) - (400x + 850y + 500) = 200x + 150y - 500$$

Again the corner points will maximize the profit, and if we try (200,0) we have a profit of $200(200) + 150(0) - 500 = \$39{,}500$, while the point (140,60) would yield a profit of $36,500 and (90,60) even less, so that the entire production should be given over to 2-cylinder engines if maximizing profit is the only consideration.

Figure 8.14

CHAPTER 8 / INEQUALITIES

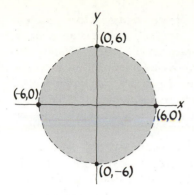

Figure 8.15

There are other inequalities in two variables whose solution sets are not bounded by straight lines. Consider, for example,

$$x^2 + y^2 < 36$$

Again, a little thought will convince you that the boundary is given by $x^2 + y^2 = 36$, which is the circle of radius 6 and with center at the origin. The plane is once again divided into two regions; now there is one inside the circle and one outside, as shown in Figure 8.14.

If we pick a point inside the circle, say $(0,0)$, we have, on substitution into our inequality, that $0^2 + 0^2 < 36$ is true and that the region inside the circle represents the solution set. The circle itself is not part of the solution set, since, say, $0^2 + 6^2 < 36$ is false and so is not a solution. We represent that by using dashes to indicate the boundary is not part of our solution. The picture looks like Figure 8.15.

EXERCISES 8.6

1. On a sheet of graph paper, draw in the line $x + y = 200$ and plot the following points. Determine which satisfy the inequality $x + y \le 200$ and which do not.
   (a) $(0,1)$ (b) $(200, -3)$ (c) $(200,3)$
   (d) $(167,33)$ (e) $(82,119)$ (f) $(-129, -81)$

2. Repeat Exercise 1 using the inequality $2x - 3y > 0$ for:
   (a) $(0,1)$ (b) $(1,0)$ (c) $(6,2)$
   (d) $(6,4)$ (e) $(-2, -1)$ (f) $(2, -7)$
   (g) All pairs in quadrant one, or two, or three, or four.

3. Draw the solution sets of the following. Indicate in each case, by a solid heavy or dotted line, whether the boundary line is included or not.
   (a) $x + y \le 6$ (b) $2x - y < -2$ (c) $3x - 7 < 4y + 5$
   (d) $2x + 6y - 5 \le 3x + 2y - 7$ (e) $x \ge 0$ (f) $y - 7 \le 3$
   (g) On the basis of a–f, find a general relation between strong inequalities and boundaries.

4. Draw pictures of the intersections of the solution sets of the following inequalities from Exercise 3. Label all corner points.
   (a) a and b (b) a and c (c) d and e
   (d) a and d and e (e) c and b and e and f

5. The following questions grow out of the engine-factory problem of the text, and are based on the original income statement of $600x + 1000y$ and the original cost statement of $400x + 850y + 500$.
   (a) Calculate the value of the income statement for each corner point and for at least five points which are not corner points.
   (b) Do Exercise (a) for the cost and profit statements. What is the minimum cost? Why is this not a practical solution to keeping costs down?
   (c) Suppose the cost of the 2-cylinder engine rises $50 per unit but the selling price cannot go up for competitive reasons. Calculate the new cost and the new profit. Find the corner point which maximizes the profit. Is anything odd about your result? Determine if any other points will maximize the profit? Explain.
   (d) Suppose we put an addition onto the factory so that the total capacity is 400 units per day. Furthermore, suppose that with the new machinery it is capable of making up to 350 two-cylinder engines and up to 150 four-cylinder engines. Everything else remains the same as in the original problem, except that the cost function becomes $400x + 825y + 3000$. Find the mix that gives the maximum income and maximum profit. What are these maximums?

6. At a given launching center, there are facilities for launching up to 70 space shots in one year. The center launches two types of satellites, manned and weather. To insure proper weather forecasts, at least 30 weather satellites must be launched. In order to keep the teams in training, at least two manned satellites must be launched, but no more than 10 of these latter can be gotten off in one year.
   (a) If it is estimated that the gain to the nation (neglecting cost) is $30 million for each weather satellite and $50 million for each manned flight, find the number of each which will maximize the gain.
   (b) The cost is $10 million for each weather satellite and $55 million for each manned satellite, plus a fixed annual charge of $100 million for running the center. Find the least cost to the taxpayer.
   (c) Find the combination of space launchings which will yield the largest net gain to the taxpayer under assumptions (a) and (b) of the problem.

7. Draw the solution sets to the following inequalities. Indicate which parts of their boundaries are included.
   (a) $x^2 + y^2 \leq 49$ (b) $x^2 + y^2 > 20$

(c) $y < x^2$

(d) $\dfrac{64}{x^2 + y^2} < 1$

(e) $xy \le 1$

(f) $4x^2 + 9y^2 \le 144$

°(g) $x^2 - 4y^2 \ge 36$

°(h) $4 \le x^2 + y^2 < 121$

°(i) $0 < x^2 + y^2 \le 1$

°(j) $0 < xy \le 1$

Chapter 9

Motion and the Beginnings of the Arithmetization of Society

9.1 Introduction

In Chapter 1 we pointed out that we all live in a society which has used mathematics to describe many of its own details. That mathematics has been successful in doing this is undeniable; the way man has developed control over his environment by use of mathematics and the consequences of mathematics bears witness to this. Whether or not this has been a good thing is not the issue; the arithmetization of our society is a fact, and it may even be the characteristic feature of modern society.

In this chapter we will explore the beginnings of arithmetization. Our story will start with Galileo, but first we shall need some background by way of contrast.

9.2 The Middle Ages

If there was one single outstanding physical problem of the Middle Ages, it was the problem of motion. How did things move, what kept them going, and why did they stop? Most especially, *why* did things move?

By the latter part of the twelfth century, most of the works of Aristotle on physics were known to Latin Europe, in part indirectly through the Arabs by way of Spain, and in part more directly from the Greeks by way of the Norman Kingdom of Sicily. Thus, by the thirteenth century, the European scholarly world was prepared for two of Aristotle's most eloquent expositors, Albertus Magnus (ca. 1200–1280) and St. Thomas Aquinas (1225–1274). It was Aquinas in particular who injected Aristotle's physics into the thinking of the Middle Ages, until this body of thought became the most widely accepted theory in the middle of the fourteenth century.

For Aristotle, rest was the natural state of any body: motion for its own sake was impossible. Motion could only exist to accomplish something—a means to an end. This in turn implied a mover capable of inducing the motion and attaining the end.

Often, the mover was obvious; for example, the horse pulling the cart, or the archer loosing the bowstring to make the arrow fly. But what of the stone falling to earth, or the spark floating upward from the fire?

Aristotle attacked this problem by dividing all motion into *natural* and *violent* motion. All matter was composed of various proportions of the four elements: earth, water, air, and fire. Generally, earth and water tended to move naturally towards the center of the worldly globe, while fire and air tended to fly upwards towards the heavens. No object was made up solely of one element, but where there was a preponderance of one or another of the elements, the natural motion of the object would be the same as that of its principal ingredient. In this way, a stone (mostly earth) would naturally fall to the ground, while a spark (chiefly fire) would naturally rise. Thus, an object's *natural tendency* could be considered to be its moving agency.

All motion which was not natural was *violent*. Since violent motion was in a direction which was different from the natural tendency, a continuing force was required to keep a body going in an unnatural direction at a given speed. Without such a force it would slow down, stop, and eventually revert to its natural direction. Of course, Aristotle recognized that, in different media, different forces would be required to move the same object at the same speed: it is much harder to push something through water than through air. The relation he arrived at was that the speed of an object was *proportional* to the force applied, and *inversely proportional* to the resistance; in modern notation,

$$S = K\left(\frac{F}{R}\right)$$

where S is the speed, F is the force, R is the resistance, and K is the constant of proportionality.

There are several interesting consequences of this formula. Note first that this implies that a constant speed requires a constant force to maintain it; and, conversely, a constant force implies that the speed is constant if all else remains the same. Secondly, observe that if the formula is correct and universal, there can be no vacuum, for if there were, any object in it would be instantly transported regardless of the force applied. This constituted one of the chief objections to the theory in the Middle Ages. If a vacuum could not exist, even God could not make one, and this represented a limitation on His powers, a theologically unacceptable idea.

There were also other objections to the Aristotelian theory: How did an arrow continue to move after it had no further contact with the bow string? There was apparently no further force being applied to it. Why did a stone fall faster and faster as it neared the ground? What increase in force was being applied to it which increased its speed?

The Aristotelians' replies to these questions constituted the weakest part of their theory. They reasoned that the arrow kept flying because of the disturbance it created in the air. That is, the air rushed in behind the arrow in order to prevent the formation of the vacuum so abhorred by nature. In this way the arrow continued to be impelled by the particles of air striking it from behind. How the air could both impede the motion (the resistance, R, in the formula $S = K(F/R)$) and supply the motivation at the same time, was a detail left to later generations.

The acceleration of the falling stone was given no better an explanation. Some reasoned that it fell faster because the increasing weight of the column of air above it applied an increase of force, while there was less resistance offered by the diminishing column of air beneath it. Others argued that it became more joyful as it neared its natural place, and therefore quickened its pace in order to arrive home more quickly!

Of course, there was more than one line of thought in those times. Not everyone who considered physics was an Aristotelian. Perhaps the most significant of the countertheories was the *Theory of Impetus,* which flourished chiefly in Paris at the end of the fourteenth century. Not surprisingly, it attacked the Aristotelians where they were weakest, at the flying arrow and the falling stone.

According to the proponents of impetus, a body such as an arrow was carried forward by an actual impetus it had acquired by being set in motion. As the arrow continued in its flight, it gradually lost its impetus, much as a red-hot iron bar loses its heat. When there was insufficient impetus to overcome its natural tendency, it fell to the ground.

For falling bodies, it was felt that as a body fell it picked up additional gravity. That is, the acceleration of a falling body was due to the effects of impetus being constantly added as a consequence of the increase in weight.

Today's scholars are not in agreement as to the extent to which this theory was a forerunner of the modern theory of momentum. It is very easy to attribute present-day thoughts to ancient scholars, to put ideas in their heads which never occurred to them. But it is just as easy to dismiss their ideas as crank, being made up only of superstitions. We can follow the line of this theory to Galileo (we can even make a good guess at the book he learned it from); and who knows, perhaps it began a train of thought leading to our theory of momentum?

EXERCISES 9.2

1. It has been said that "A universe constructed on the mechanics of Aristotle had the door halfway open to the spirits . . ." Explain what was meant by this.

*2. What is the chief difference between the Theory of Impetus and the modern theory of momentum, in the case of the arrow? Modern theory defines momentum to be the product of the *mass times velocity.*

9.3 Galileo Galilei

Historically, it is seldom that we can point to one man and say with any degree of surety that *he* was the founder of this or that great branch of knowledge. For the most part, the great theories in our heritage have gradually developed and evolved with time and in response to certain problems. But if any fundamental study has had a father, surely Galileo Galilei (1564–1642), the Italian, was the sire of modern science, or, at the very least, of modern physics.

This is not to say that every idea which came from Galileo was basic to physics, nor were even the essential ideas all fully formulated and correct down to the last and final detail. Many of his concepts were quite primitive and incomplete, when considered from a modern point of view.

What Galileo did do was to promulgate standards by which good science can be judged, as well as to lay out the very basic tenents of scientific research. In particular, he broke sharply with the Aristotelian past of his predecessors. Nature was to be Galileo's teacher; authority was to be ignored to the extent that it did not agree with nature.

For Galileo, mathematics was to be the basis for science in two ways. First, mathematics (in his day, essentially Euclidean geometry and arithmetic) was to be the model for scientific inquiry. A few basic assumptions should be hypothesized and, if possible, verified by carefully conceived experiments; these corresponded to the axioms of geometry. From these assumptions, various properties would be deduced about the behavior of objects; these were to be the theorems of physics. These, too, were to be compared with reality by means of experiments, so that theory should follow nature.

But this was not all. Mathematical objects would provide the model for the objects of physics. That is, the laws he deduced from his assumptions were not laws about real objects in the real world, but abstract objects interacting in an abstract universe. Thus, the behavior of falling objects was to be studied in a world without air resistance; and the motion of objects on the surface of the earth was to be studied by considering perfect spheres rolling about on a plane. This meant that the results of the experiments could not be expected to agree exactly with the theory, but the differences should be small and explainable.

Secondly, mathematics was to provide the language of physical laws; that is, the results of his studies would be presented as mathematical expressions. In this way, Galileo shifted the focus of attention of physics from the *qualities* of matter (gravity, levity, color, texture) to *quantities*, things which could be described mathematically. The study of causes was not for him.

There was one other important break that Galileo made with the past: he would only explore the universe a little at a time. Galileo's predecessors had all been prone to propose vast general theories which were meant to explain all natural phenomena; the details were then expected to fall into place in accord with the overall plan. Galileo, on the other hand, felt that, before a general theory could be promulgated, individual details would have to be understood.

Galileo based his study of motion on the revolutionary hypothesis that a *constant force* would produce a *constant acceleration, rather than a constant velocity*, as the Aristotelians felt. That is, Galileo supposed that, if a uniform force were applied to an object, its velocity would show the same increase over the same time periods, and the increase would be proportional to the time interval.

For example, if an object subjected to a constant force accelerated from rest to 12 feet per second in a period of one second, then after two seconds it would be traveling at 24 feet per second, while after 6 seconds it would be moving at 72 feet per second. Also, after $\frac{1}{3}$ of a second it would be going at 4 feet per second. In modern notation, its velocity, v, would be given by the formula $v = 12t$, where v is measured in feet per second and t is time measured in seconds.

Note that Galileo's hypothesis eliminates the concept of the natural state of an object. The inanimate things of this universe are no longer endowed with a "desire" to go places and do things on their own. Instead, an object at rest must be pushed (or, more precisely, have a force somehow applied to it) in order to set it in motion; and it will not change its speed unless force is applied. Galileo understood this principle, called the *principle of inertia*, well enough for his purposes, although his formulation of it was incomplete. A refined statement was made by Newton almost half a century later, and we will give his formulation of the inertial principle in the next section when we study the modern theory.

Once Galileo was able to show that the velocity of an object being forced at a constant rate from rest was proportional to the elapsed time, he used this to show that the distance traveled by the same object was proportional to the square of the elapsed time. That is, in two seconds it would travel four times as far as in the first second; in three seconds, nine times as far; and so on.

Thus, our earlier particle whose velocity behaved according to the formula $v = 12t$, would obey the formula for distance by $d = 6t^2$, where the distance d is to be measured in feet. (Note that the constant of proportionality is half of the earlier one for velocity: this is no accident, but can be shown to be quite general.) Thus, our particle will go 6 feet in the first second, 24 feet in the first two seconds, and 54 feet after three seconds.

The latter result was one which Galileo could test by experimentation. He built a board that had a smooth groove and that could be tilted at various angles. A ball was rolled down the groove, the constant force applied to it being the "diluted" force of gravity. Note this extremely important supposition: *The force due to gravitation was constant.*

Galileo encountered a good deal of difficulty in timing his experiments, as there existed no accurate clock. Galileo hit on the idea of using water; he constructed a large container with a very small tube near its base. He could then time his experiments by weighing the amount of water which flowed out of his apparatus.

Galileo's experiments brilliantly confirmed his theories. No matter how steeply he pitched the board, the distance traveled by the ball rolling down it was always proportional to the square of the time. Of course, the constant of proportionality changed with the steepness, but the basic law held in all cases.

Then Galileo took one of those mental leaps which characterize the genius and seem so easy and obvious *after* they have been made. Since the distance was always proportional to the square of the time *no matter what the angle of the plane*, this same law would hold when it was almost vertical; and hence, by extension, when it *was* vertical. That is, a freely falling body would also have the property that the *length of the fall varied directly as the square of time of the fall.*

Galileo also observed in his experiments that the distance did not depend on the weight of the ball. From this he reasoned that the distance an object would travel in free fall would not depend upon its weight. He also knew about the results of an earlier experiment performed in Pisa by Vincenzio Renieri, who had dropped balls of unequal weight from the Leaning Tower.[1] He found that it was true that the heavier reached the ground sooner, but only very slightly so. Galileo assigned the difference to the resistance of the air. He claimed that in a vacuum there would be no difference at all.

Galileo himself conceived of an experiment that would show that the speed of freely falling bodies is independent of weight. The difficulty was that in order to show a measurable difference between the speeds of bodies of different weights, they would have to be dropped from very great heights. Galileo saw that it was possible to amplify the differences, if any, by a pendulum effect. If two different weights were fixed to threads of the same length and if they were dropped from the same height (as in Figure 9.1), any real difference in the speed would become more apparent as the apparatus oscillated back and forth, multiplying and exaggerating the effect. In particular, the one

---

[1]It is extremely unlikely that Galileo performed this experiment, although tradition says he did. No record of it has ever been found either in the archives of Pisa or in Galileo's own papers.

Figure 9.1

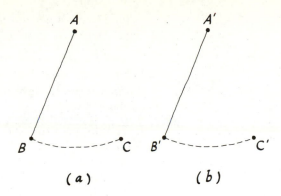

(a)   (b)

with the lighter bob (weight) would move more slowly than the heavier if the Aristotelians were correct. In fact, if the heavier were ten times the weight of the lighter, then it would be ten times as fast, according to them. Galileo found no appreciable difference between the time of one weight or another in his experiments.

Galileo's contributions to science were not only in the field of falling bodies. He made other discoveries in mechanics, including the first correct calculation of the path of a cannonball. He also invented telescopic astronomy, although not the telescope itself. He was the first to see Jupiter's moons, the phases of Venus, and the craters of the moon, as well as sunspots.

EXERCISES 9.3

1. It has been said that the Aristotelians made errors in their physics because they observed nature too carefully, but not with enough attention. When considering the motion of a cart, it was necessary to apply a force just to keep it going at the same speed. Explain this apparent exception to Galileo's inertial principle.

2. Consider a body which accelerates from rest to 10 feet per second in one second.
   (a) How fast is it going after two seconds? after four seconds? after $\frac{1}{3}$ second?
   (b) How far does it travel in two seconds? in four seconds? in $\frac{1}{3}$ second?

(c) Derive a general formula for its velocity after t seconds.

(d) Derive a general formula for the distance traveled after t seconds.

3. Suppose a body has accelerated from rest to 24 feet per second in three seconds. Answer questions (a), (b), (c), and (d) of Exercise 2 about it.

4. The distance traveled by our particle of the text in the $(t + 1)^{\text{th}}$ second [that is, between time t and time $(t + 1)$] is given by $6(2t + 1)$.

(a) Verify this from the example in the text.

(b) Derive it algebraically. [*Hint*: By the end of the $(t + 1)^{\text{th}}$ second, the particle has gone $6(t + 1)^2$ feet.]

5. Why was it impossible for Galileo to directly verify the law that "distance is proportional to the square of the time" for freely falling bodies?

6. Consider Galileo's interpretation of the Leaning Tower experiment.

(a) Why would air resistance have more relative effect on a lighter ball than a heavier? (*Hint*: Carefully measure the diameters of oranges of differing weights.)

(b) Where was Galileo using the mathematics-like abstraction process?

7. Galileo also reasoned, in the pendulum experiment, that if the bob (weight) were lifted to B, then the point of furthest rise, C, would lie on a horizontal line from B. (See Figure 9.1) That is, it would rise to exactly the same height as it had been lifted.

(a) It did not, in practice. There are at least three reasons why the real pendulum would not behave like the ideal one. Find at least two of them.

(b) If the real pendulum had behaved like the ideal one, would the motion be perpetual? Explain.

*8. Those familiar with Newton's law of gravity know that the force of gravitational attraction between two objects is

$$F = K\left(\frac{mM}{r^2}\right)$$

where m and M are the masses of the objects and r is the distance between their centers; K is a constant of proportionality. Thus the gravitational pull on an object *does* depend on its mass as well as on the height from which it is dropped. How do you explain this contradiction with Galileo's experimental results?

9.4 The Modern Theory of Falling Objects

Modern theory of motion is based on three axioms. They were first proposed in their modern form by the English mathematician, physicist, and theologian, Sir Isaac Newton (1642–1727), who was born, by odd coincidence, in the same year that Galileo died.

Newton's mechanics, which includes the theory of freely falling bodies, is based on the three axioms that have played as an important a role in our civilization as the ten axioms of Euclid, though they are considerably less well-known. Somewhat restated, they are:

AXIOM 9.1   *An object at rest tends to remain at rest. An object in motion tends to continue in motion in the same straight line and at the same rate of speed.*

This is the inertial principle we first encountered in a primitive form as expounded by Galileo. It means that an object at rest must have a force applied to it to get it going. Also, a body in motion must be forced if it is to change its speed or its direction. In particular, an object cannot be slowed unless there is an action causing it to do so.

AXIOM 9.2   *The acceleration of a body is proportional to the force applied to it and is in a linear direction, the direction in which the force is applied.*

The first part of this axiom is usually abbreviated with the formula $F = ma$, where F is the force, a is the acceleration, and m is the mass. It can be looked at as the definition of *mass* in terms of force and acceleration. Here we must distinguish carefully between mass and weight. On earth the two tend to become confused because near the surface of the earth they are essentially proportional.

Weight is the measure of the gravitational attraction between an object and the earth, while mass, as we can see by Axiom 9.2, is an object's resistance to being pushed. For example, suppose we are in the outer reaches of space where there is essentially no gravitational force. Any object we have with us will be weightless, although it will still have the same mass as it did here on earth. A small ball of foil wrap will be about 3000 times easier to set into motion, say, than a man who weighed 200 pounds on earth, even though in space both will weigh the same—nothing!

Note that if we apply the same force to both the man and the foil, the foil would accelerate 3000 times faster than the man. Thus, if the man were accelerated at $\frac{1}{10}$ foot per second per second[1] by a certain force, then the foil would be accelerated at 300 feet per second per second by the same force.

AXIOM 9.3   *For every force applied to one object by a second in a given direction, there is an equal force applied by the first to the second in an opposite direction.*

This has often been abbreviated to the statement, "For every action there is an equal and opposite reaction." Anyone who has ever fired a rifle or pistol has felt this axiom at work, since it explains the recoil of such weapons.

But there are other, less warlike, examples. Suppose we direct a stream of water at a basketball which has been placed on a flat surface (not the floor of your gym!). The water is diverted from its straight path by the ball, showing that the ball has exerted some force on it (see Axiom 9.1). In addition, the ball starts to move, showing that the water is exerting a force on the ball (also Axiom 9.1). Note that the ball moves in the same direction as the mainstream, while most of the splashes bounce straight back (except for those drops influenced by the curvature of the ball).

Suppose we return to outer space and our man playing with the ball of foil. If he flicks it out in front of him so that it is accelerating at 300 feet per second per second, that act will push him backwards at $\frac{1}{10}$ of a foot per second per second.

If we apply all of this, we see that Galileo's results are included in the theory of Newton. The force due to gravity is essentially constant near the earth's surface; this means that the balance of his deductions follow as he reasoned them. In our system of feet and seconds, the acceleration due to the force of gravity is 32 feet per second per second, so that the velocity, v, of a freely falling object (starting from rest) after t seconds is given by $v = 32t$, and the distance, d, fallen after t seconds is $d = 16t^2$.

We can illustrate the formula by the following idea. Suppose we draw a right triangle (see Figure 9.2) with base t representing the time

---

[1]This peculiar looking system of units (feet per second per second) makes perfectly good sense for acceleration. In this case it means that the velocity is increased by $\frac{1}{10}$ of a foot per second, *in each second*.

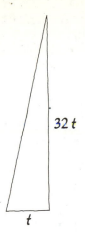

$32\,t$

t

Figure 9.2

k

t

Figure 9.3

a particle is in free-falling motion, and height $32t$ representing the velocity it attains in this time; then the area of this triangle will be $16t^2$ (by the familiar formula $A = \frac{1}{2}bh$). Note that this result holds for any t.

Now consider a particle which has constant velocity k. Then the distance it travels in t seconds is kt, representable by the area of a rectangle with base t and height k (Figure 9.3).

Observe that the distance traveled is the area of a geometric figure constructible from the data. This identification of the distance traveled with the areas of geometric figures is not an accident, but can be justified with advanced techniques. Its use goes back at least to Galileo.

Next, suppose we are standing on top of a building and throw an object straight down with a speed of, say, 48 feet per second. Then, since the force due to gravity is acting in the same direction as the force of the throw, it will augment the speed of the object. That is (according to Axiom 9.2), the speed at any time will be the same as if the sum of the two forces were acting on it. But, by Axiom 9.1, the speed due to the throw would remain constant at the original 48 feet per second, as no further force is provided by the throwing arm once contact has been lost; also by the earlier theory, the speed due to gravity is $32t$, so that the sum is $(32t + 48)$ feet per second.

It seems reasonable to calculate that the distance traveled also adds up the same way, so if we draw a geometric figure (see Figure 9.4) corresponding to the velocity of $(32t + 48)$ feet per second, we have a right trapezoid of area $48t + 16t^2$, and, reasoning by analogy, we have that this is the formula for distance. This formula has been verified experimentally. In a similar way, we can show that any object thrown downward with an initial speed k will have velocity $(32t + k)$, and the distance it will travel will be $16t^2 + kt$ after t seconds.

Next, suppose we are standing on the ground and throw an object straight *up*. What formula can we derive to describe its subsequent motion?

Figure 9.4

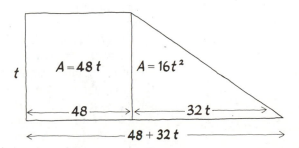

t

$A = 48\,t$ $A = 16t^2$

48 $32\,t$

$48 + 32\,t$

Before we begin this task, we must carefully distinguish between velocity and speed. Up to now we have used the terms interchangeably because all our motion has been in one direction only. However, now we are going to have to deal with change of direction. *Velocity* measures rate of change of position with respect to time, where the *direction is taken into account* by its sign. *Speed* measures the same thing, *without regard to direction.* Typical of the units used are miles per hour and feet per second. We observe that velocity can be either positive or negative, but speed is *never* negative. Strictly speaking, the speed–velocity formulas we have derived up to now have actually been for velocity.

To see the difference, consider two cars which are going in opposite directions, on a flat, straight highway. If they are both going at 50 miles per hour, then that is the speed of each. But since they are moving in opposite directions, one will have a velocity of 50 miles per hour, while the other's velocity will be −50 miles per hour. Which one has the positive velocity and which the negative? This is a matter which depends only on the choice of the coordinate systems.

To return to our problem, suppose we throw a ball straight up into the air at 128 feet per second. Since we know from experience that it will change its direction and fall at our feet, we must pick one direction as positive, in which case the other automatically becomes negative. The one we choose is a matter of taste and convenience alone, but since most of us are used to considering up as positive, we shall pick our positive direction to correspond with this prejudice.

In this case, the 128-feet-per-second throw we have made is in the positive direction. Once the ball has left our hand, we are applying no further force to it, so that it would tend to fly upward at 128 feet per second indefinitely if it weren't for the force of gravity acting upon it in an opposite direction. Notice that this means it is slowing the ball down, and so we indicate the acceleration due to gravity as −32. If gravity alone were acting on the ball, we have seen that its velocity would be −32t (don't forget that falling is now in a negative direction), and if we again assume that velocity is additive, then we arrive at the formula $v = -32t + 128$ for the velocity of the ball after t seconds.

We could next ask ourselves how long the ball will continue to rise. In terms of our coordinate system, this is the same as asking "When will the velocity cease to be positive?" Clearly, this will occur when the velocity, v, is zero; and thus, setting $-32t + 128$ equal to zero and solving for t, we find that it will reach its maximum height four seconds later, since it rises no further.

To calculate the distance it rises in time t, observe that any formula we derive will be valid only for t between zero and four. Now, in calculating the distance traveled, it will be distance covered by an object moving at a constant 128 feet per second ($128t$) less the distance lost by the slowing up caused by gravity. But the loss is the same as the distance covered by a freely falling object, $16t^2$. This gives us a net distance covered of $d = 128t - 16t^2$; observe that this formula is valid so far only for t between zero and four. In particular, it gives us a maximum distance (at $t = 4$, when $v = 0$) of $128 \cdot 4 - 16 \cdot 4^2 = 256$ feet.

We can draw several interesting conclusions from all of this. We note that the ball is at rest momentarily at a height of 256 feet above the ground, and we can now treat it as a body freely falling from rest. Picking a new variable T (since t has been used already in this problem), which is zero when the ball is at the top of its trajectory, our earlier work tells us that its velocity is $-32T$ and the distance traveled is $-16T^2$. (Again, recall the minus signs are necessary because up is positive.)

Now we know the total distance traveled downward, -256 feet, so that we have $-16T^2 = -256$ or $T^2 = 16$ or $T = 4$. Thus, we have the not very surprising result that it takes as long to fall back to earth as it did to rise in the first place. Furthermore, since $T = 4$, the ball hits the ground with a velocity of $-32 \times 4 = -128$ feet per second, a speed equal to but opposite in direction to the initial velocity it was given.

We can also show that the expression $d = 128t - 16t^2$ can be extended to the entire time the ball is in the air. To see how this is done, observe that the ball is at a distance of 256 feet above the ground when it begins its descent. Thus, its distance at time T is $d = 256 - 16T^2$, since it will have fallen $-16T^2$ feet. But when $T = 0$, $t = 4$, so $T = t - 4$ so that substituting for T in d gives $d = 256 - 16(t - 4)^2$ or

$$d = 256 - 16(t^2 - 8t + 16) = 128t - 16t^2$$

Furthermore, the method is quite general. That is, if we throw a ball in the air with a velocity of k feet per second, then the distance above the ground, d, after t seconds, will be given by $d = kt - 16t^2$, which will be valid for those values of t which yield a nonnegative d.

EXERCISES 9.4

1. In some of the early space walks taken by the American astronauts, they used a gadget like a water pistol to move around in space. Why did it work?

2. In view of Axiom 9.1, how do you explain the fact that we must keep feeding gas (applying force) to the engine of our automobile just to maintain a given speed on a level road?

3. Assuming that the force due to gravity on the moon is $\frac{1}{6}$ that of earth,
   (a) If an object is dropped near the surface of the moon, how fast will it be going after one second?
   (b) Find the general formula for the velocity after t seconds.

4. If a ball is thrown downward from the top of a 1000-foot building with an initial speed of 48 feet per second, how far will it have gone:
   (a) In 1 second?   (b) In 3 seconds?
   (c) In 10 seconds?

5. Suppose an object is thrown downward at an initial speed of 16 feet per second from a cliff which is 1760 feet high.
   (a) How fast will it be going after 1 second?
   (b) Find the formula for its velocity after t seconds.
   (c) Find the formula for the distance it has fallen after t seconds.
   (d) When will it hit the bottom of the cliff?

6. Consider the problem in the text where a ball was thrown upward with an initial velocity of 128 feet per second.
   (a) What was its velocity 2 seconds after it was thrown? 3 seconds? 7 seconds? 9 seconds?
   (b) What was its altitude above the ground after 2 seconds? 3 seconds? 7 seconds? 9 seconds?
   (c) Notice that $d = 0$ when $t = 0$ or $t = 8$. What is the significance of these values of t?

7. Suppose a ball was thrown straight up with an initial velocity of 320 feet per second.
   (a) How long will it continue to rise?
   (b) How high will it go?
   (c) What is its velocity when it hits the ground? Is there any general conclusion you can arrive at inductively?

8. Suppose a feather is thrown up with an initial velocity of 320 feet per second. Can you answer the questions in Exercise 7? Why? What has been assumed throughout the entire discussion?

9.5 Combination of Two Motions at Once

In continuing our study of motion, suppose we now are standing on a high vertical cliff and throw a ball out straight in front of us. From experience we know that the ball will travel gracefully in an arc, land-

Figure 9.5

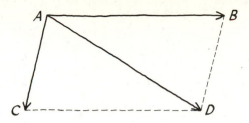

ing some distance from the base. But how far from the base? How long will it be in the air? What is the shape of the arc?

To treat this and similiar questions, we must go back to Axiom 9.2 and consider what happens if an object is acted upon by two forces at the same time. It is implicit in this axiom that the velocity imparted by one of the forces is independent of the velocity imparted by the other. Furthermore, the combined effect on the object obeys the *Law of Parallelogram of Forces* as follows: If the action of one force would give sufficient velocity to move a particle from A to B in a certain period of time, while the other would move it from A to C in the same period, then the combined effort of both forces would move it from A to D in that time, where D is the fourth point of the parallelogram with sides AB and AC. (See Figure 9.5.)

To see this, observe that the first force would cause the particle to move to the left a distance AB and that, according to Axiom 9.2, the amount of this displacement is not affected by the second force. Similarly, the length of the second motion is unaffected by the first, and must be equal to AC. To put it differently, the first force must push the particle to some point on the line BD in the time period (by Axiom 9.2), while the second will carry to some point on CD. Hence, the *resultant* of the two together will transport the particle to the *intersection* of the two lines, namely, D.

Consider, now, what happens when we stand on some high place and throw a ball straight out in front of us. Just as it is released there are two forces acting upon it. There is a horizontal force imparted by the force of the throw, and a vertical force, downward, the force of gravity. Suppose we throw a ball straight out with a velocity of 25 feet per second, as shown in Figure 9.6.

Since no further *horizontal* force is being applied (we ignore air resistance), the velocity will remain constant, and the distance traveled horizontally in t seconds will be $25t$ feet. Simultaneously, the pull of gravity will (according to our previous work and the independence principle) force our particle downward a distance of $16t^2$ feet.

If we now impose a coordinate system, with the origin at the point where we are standing, as shown in Figure 9.7, we see that, at any time $t \geq 0$, the x coordinate will be $25t$ while the y coordinate will be $-16t^2$. This will be valid until the ball hits the ground.

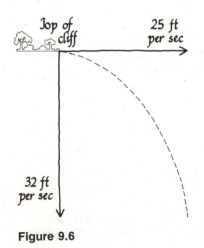

Figure 9.6

Figure 9.7

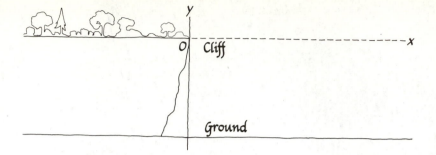

In order to find the shape of the curve described by the ball, we note that if $x = 25t$, then $x^2 = 625t^2$ or $t^2 = x^2/625$. Substituting for t^2 in the expression for y, we have that

$$y = \left(\frac{-16}{625}\right)x^2$$

which is the formula for an arc of a parabola. The path will look like Figure 9.8.

The method is perfectly general. If we throw a ball off a cliff with a velocity of k feet per second, then it will travel horizontally kt feet in t seconds, while it will drop $16t^2$ feet in the same time. (Of course, the force due to gravity does not change.) This gives us the equation for the path of the ball:

$$y = \frac{-16x^2}{k^2}$$

Finally, suppose we stand on the ground and throw a ball on an angle away from us. Once again we must appeal to the principle of independence, but this time we turn it around.

To be specific, suppose we throw a ball away at a 30-degree angle with a velocity of 50 feet per second. That means that in t seconds it will travel $50t$ feet at 30 degrees from the horizontal like Figure 9.9.

Figure 9.8

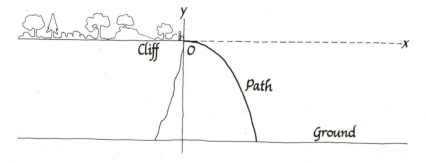

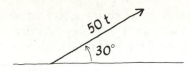

Figure 9.9

We can now use the law of the parallelogram of forces (backwards, as it were) by considering our force to be the resultant of two independent forces, one horizontal and the other vertical. We could have picked almost any other directions to decompose our original force into, but these are the most convenient, since we know about objects traveling in these directions.

Putting in the component forces in our diagram, we get (from trigonometry) that the horizontal component is $(\cos 30°)50t$, while the vertical is $(\sin 30°)50t$. (See Figure 9.10.) Since $\cos 30° = \dfrac{\sqrt{3}}{2}$ and $\sin 30° = \frac{1}{2}$, we have that the horizontal component is

$$\frac{(\sqrt{3})}{2}50t = 25\sqrt{3}t$$

while the vertical component is $25t$. To put it in another way, throwing a ball into the air at 50 feet per second at a 30° angle is the same thing as hitting it simultaneously with two forces, one which would move it horizontally $25\sqrt{3}t$ feet in t seconds, while the other would push it straight up at a rate of 25 feet per second.

We know from earlier work in this chapter that the height above the ground after t seconds will be $25t - 16t^2$. If we now impose a coordinate system with the origin at the point of the throw, this tells us that the y coordinate will be

$$y = 25t - 16t^2$$

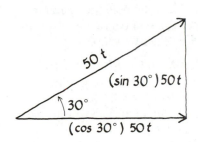

Figure 9.10

Just as in the cliff problem, the x coordinate will be $x = 25\sqrt{3}t$. Thus $t = x/(25\sqrt{3})$ and $t^2 = x^2/1875$. Substituting for t and t^2 in the expression for y we get

$$y = \frac{x}{\sqrt{3}} - \frac{16x^2}{1875}$$

This is valid for all values of x which make y nonnegative. Once again we have the arc of a parabola; this time the arc will look like Figure 9.11.

Figure 9.11

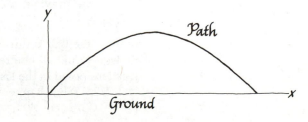

Again this method is perfectly general. If we throw a ball at an angle Θ with a velocity k, then after t seconds it will have traveled horizontally $(\cos \Theta)kt$ feet and vertically $(\sin \Theta)kt - 16t^2$ feet, and the resulting arc is parabolic.

EXERCISES 9.5

1. If I throw a ball straight off a building and at the same instant drop another, which will hit the ground first? Why?

2. Suppose I drop a ski pole from a high moving chair-lift just as my chair is passing a supporting pylon. When the pole hits the ground will it hit at the base of the pylon or directly below my chair, assuming the chair does not change its velocity? Why?

3. In the example in the text, suppose the cliff is 576 feet high.
   (a) How long is the ball in the air before it hits the ground?
   (b) How far from the foot of the cliff does it fall?
   (c) What assumptions have you made about the geography near the cliff in your solutions in (a) and (b)?
   (d) For what values of x is $y = (-16/625)x^2$ a reasonable description of the path of the ball?

4. (a) If a ball is thrown off of a 576-foot cliff, how fast must it be thrown to hit 100 feet from the base?
   *(b) Is it ever possible to throw it (or have it propelled) hard enough so that it will never hit the ground? Does your answer have anything to do with Exercise 3 (c)?

5. Consider the problem in the text, of the ball thrown at an angle of 30 degrees.
   (a) How long will it be in the air?
   (b) How far horizontally, will it travel?
   (c) Use the previous section to find out the highest point it will reach.

6. Suppose the ball is thrown up at an angle of 45 degrees instead of 30 degrees and all else remains the same as in the text. Answer the questions posed in the last exercise. (*Hint:* $\sin 45° = 1/\sqrt{2}$, $\cos 45° = 1/\sqrt{2}$.) What happens if the angle of throw is 60 degrees ($\sin 60° = \sqrt{3}/2$, $\cos 60° = \frac{1}{2}$)?

REFERENCES

Bell, Arthur E. *Newtonian Science*. Edward Arnold (Publishers) Ltd., London, 1960.

Brewster, David. *Memoirs of the Life, Writings and Discoveries of Sir Isaac Newton*. 2 vols. Johnson Reprint Corporation, New York, 1965.

Brodetsky, S. *Sir Isaac Newton*. Luce, Boston, 1927.

Copernicus. *On the Revolutions of the Heavenly Spheres*. Translated by C. G. Waths. Great Books of the Western World, vol. 16. Encyclopaedia Britannica, Inc., Chicago, 1952.

De Santillana, Giorgi. *The Crime of Galileo*. University of Chicago Press, Chicago, 1955.

Drake, Stillman, ed. and trans. *Discoveries and Opinions of Galileo*. Doubleday Publishing Company, Anchor Books, Garden City, N.Y., 1957.

Drake, Stillman. *Galileo Studies, Personality, Tradition, and Revolution*. The University of Michigan Press, Ann Arbor, 1970.

Easley, J. A., Jr., and Maurice M. Tatsuoka. *Scientific Thought: Cases from Classical Physics*. Allyn & Bacon, Inc., Boston, 1968.

Kline, Morris. *Mathematics, A Cultural Approach*. Addison-Wesley Publishing Co., Inc., Reading, Mass., 1962.

Kline, Morris. *Mathematics and the Physical World*. Doubleday Publishing Company, Anchor Books, Garden City, N.Y., 1963.

Mach, Ernst. *The Science of Mechanics*. Translated by T. J. McCormack. Open Court Publishing Company, London, 1942.

Manuel, Frank E. *A Portrait of Isaac Newton*. Harvard University Press, Cambridge, 1968.

Maxwell, James C. *Matter and Motion*. 1877. Reprint. Dover Publications, Inc., New York.

Namer, Emile. *Galileo, Searcher of the Heavens*. Translated by S. Harris. Robert M. McBride Co., Inc., New York, 1931.

Newman, James R. *The World of Mathematics*. 4 vols. Simon & Schuster, Inc., New York, 1956.

Newton, Isaac. *Mathematical Principles of Natural Philosophy*. Translated by A. Motte and revised by F. Cajori. Great Books of the Western World, vol. 34. Encyclopaedia Britannica, Inc., Chicago, 1952.

North, J. D. *Isaac Newton*. Oxford University Press, London, 1967.

Ptolemy. *The Almagest*. Translated by R. Catesby Taliaferro. Great Books of the Western World, vol. 16. Encyclopaedia Britannica, Inc., Chicago, 1952.

Renyi, Alfred. *Dialogues on Mathematics*. Holden-Day, Inc., San Francisco, 1967.

Seeger, Raymond J. *Galileo Galilei, His Life and His Works*. Pergamon Press, New York, 1966.

Sutton, O. G. *Mathematics in Action.* Bell & Sons, Ltd., London, 1958.

Whitehead, Alfred North. *The Interpretation of Science.* Edited by A. H. Johnson. The Bobbs-Merrill Co., Inc. New York, 1961.

Whitehead, Alfred North. *An Introduction to Mathematics.* Oxford University Press, New York, 1958.

<table>
<tr><td>

Chapter 10

</td><td>

</td></tr>
</table>

Calculus:
A Sampler

10.1 Introduction

When you studied algebra in high school and learned to solve word problems you may have noticed a peculiarity associated with these problems; in problems involving change, particularly motion, they weren't very lifelike; they weren't really very good models of the real world.

When you studied the movements of a car going from A to B, it always moved at a *constant* velocity. The men who were digging the ditch always worked at a *constant* rate. The man who has been endlessly rowing up and down that river has been forever rowing at a *constant* rate, and even the river current flowed along without change.

Now real objects don't behave that way. Automobiles speed up and slow down, men tire when digging ditches or when rowing, and a river is full of swirls and eddies so that the current varies considerably from one point to the next. These are situations which cannot be treated by algebra alone; the algebraic machinery just cannot handle it.

What is needed is a new mathematics, a mathematics that can treat change realistically. Calculus was developed in response to this need; that is, calculus is the mathematics of change.

Note that this does not mean that algebra serves no practical purpose. There are many useful problems which are quite suited to algebraic solutions, problems involving static or one-time-only situations. Also, algebra provides much of the language used in calculus, where algebra's ability to express complicated ideas concisely and unambiguously are very valuable.

Before we proceed with our study of calculus, we are going to look at two problems which will serve to focus our attention on the new concepts we will need.

10.2 Two Problems of Definition

In the previous chapter, when we discussed velocity, we were actually appealing to an undefined notion of considerable complexity. To be more precise, we developed a theory which established that the velocity of a ball (whatever we meant by that) changed from moment to moment, under the constant acceleration due to gravity. In fact, for a freely falling body, we were able to derive the formula $v = 32t$ for velocity, showing how it varied with time. Thus, it makes perfectly good sense to ask what we mean by instantaneous velocity beyond our intuitive feelings; that is, precisely what is velocity at a given instant in time?

To answer this, we shall start with an easier idea, *average velocity* along a straight line. To fix our ideas, suppose we start with an object moving along a straight line, and then impose a coordinate system on it, so that it becomes a number line. Then its *average velocity* over a given time period is defined to be the *position where it was at the end of the period minus its position at the beginning of the period, all divided by length of the time period*, that is, the *distance covered divided by the elapsed time.*

For example, if we started at a spot -2 units from zero and 6 seconds later wound up 5 units to the right as in this diagram,

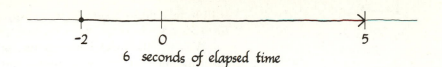

6 *seconds of elapsed time*

then the average velocity would be

$$\frac{5 - (-2)}{6} = \frac{7}{6} \quad \text{units per second}$$

Very often our position will be given as an algebraic expression of time. We call such an expression a *position function*. In the case of the freely falling body, we had that $d = 16t^2$, indicating that the body was $16t^2$ units from where it had been released. In another case, we had a position function $d = 128t - 16t^2$.

Let us consider some average velocities for the distance function $d = 128t - 16t^2$. Between $t = 0$ and $t = 4$, we get

$$v_{av} = \frac{128 \cdot 4 - 16 \cdot 4^2 - 0}{4}$$

$$= \frac{512 - 256}{4} = 64 \quad \text{feet per second}$$

Similarly, between $t = 1$ and $t = 6$ we have

$$v_{av} = \frac{(128 \cdot 6 - 16 \cdot 36) - (128 - 16)}{5}$$

$$= \frac{(768 - 576) - (112)}{5}$$

$$= \frac{192 - 112}{5} = \frac{80}{5} = 14 \quad \text{feet per second}$$

and between $t = 3$ and $t = 8$ we have

$$\frac{(128 \cdot 8 - 16 \cdot 64) - (128 \cdot 3 - 16 \cdot 9)}{5}$$

$$= \frac{0 - (240)}{5} = -48 \quad \text{feet per second}$$

Now these last two results, in particular, are not very illuminating. We know from our work in the last chapter that the ball rose until $t = 4$ and fell thereafter. In terms of velocity and the coordinates picked for this problem, this means that the (instantaneous) velocity (still to be defined precisely) was positive for $0 < t < 4$ and negative for $4 < t < 8$.

We can get a much better picture of what is going on if we divide the time interval into small pieces and look at the average velocity in each piece.

Thus, we might split the time interval between $t = 1$ and $t = 6$ into the intervals $t = 1$ to $t = 4$ and $t = 4$ to $t = 6$. In these cases, the average velocities work out to be 48 feet per second and -32 feet per second. At least now our average velocities have the ball going in the right direction. To obtain an even better analysis of the ball's velocity we would have to keep dividing up the interval into many different subintervals.

Suppose we denote one of these intervals by $(t, t + \Delta t)$. (The symbol Δt is read "delta t"; the upper-case delta, Δ, is an almost universal symbol meaning "change in.") Then the average velocity for our particular position function in this interval is

$$\frac{[128(t + \Delta t) - 16(t + \Delta t)^2] - (128t - 16t^2)}{t + \Delta t - t}$$

To see where this expression comes from, note that the part in square brackets is the value of the position function at time $t + \Delta t$; that is, it represents the position of the particle on the number line at the end of the interval. The expression in the round brackets is the position of the particle at the beginning of the time interval. The denominator, $t + \Delta t - t$, is merely the ending time, $t + \Delta t$, minus the starting time.

The next step is to simplify; multiplying out everything in the *numerator*, we get

$$\cancel{128t} + 128\Delta t - \cancel{16t^2} - 32t\Delta t - 16(\Delta t)^2 - \cancel{128t} + \cancel{16t^2}$$
$$= 128\Delta t - 32\Delta t - 16(\Delta t)^2$$
$$= \Delta t(128 - 32t - 16\Delta t)$$

The last step comes from the fact that Δt is a factor.

The denominator simplifies into

$$\cancel{t} + \Delta t - \cancel{t} = \Delta t$$

so the whole quotient reduces to

$$\frac{\Delta t(128 - 32t - 16\Delta t)}{\Delta t}$$

Observe that since Δt *is not zero*, we can divide numerator and denominator by Δt, and get that the average velocity in the interval $(t, t + \Delta t)$ is

$$v_{av} = 128 - 32t - 16\Delta t$$

From this we see that, as long as the position function d makes sense for all the values in the interval $(t, t, + \Delta t)$, the average velocity will be given by the formula just derived, which is valid for all $\Delta t \neq 0$.

For example, for t between 1 and 4, we have $t = 1$ and $\Delta t = 3$ in our formula. Then the average velocity is

$$128 - 32 \cdot 1 - 16 \cdot 3 = 128 - 32 - 48 = 48$$

and between $t = 4$ and 6, we have $\Delta t = 2$, and the average velocity is

$$128 - 32 \cdot 4 - 16 \cdot 2 = -32$$

both agreeing with our previous results.

Something interesting happens to the formula $128 - 32t - 16\Delta t$ as Δt gets close to zero — or, what is the same thing, as the length of the subinterval gets smaller. The average velocity gets nearer to $128 - 32t$. (This is easy to see because $16\Delta t$ will get very small as Δt gets very small.) We can say that $128 - 32t - 16\Delta t$ closes in on $128 - 32t$ as Δt closes in on zero. We write this as

$$\lim_{\Delta t \to 0} (128 - 32t - 16\Delta t) = 128 - 32t$$

This is read "the limit as Δt goes to zero of $128 - 32t - 16\Delta t$ is $128 - 32t$."

The fact that at $\Delta t = 0$, the formula $128 - 32t - 16\Delta t$ also takes on the value $128 - 32t$ is interesting, but irrelevant for our purposes. Why? Because, recall, that the formula represents average velocity only when $\Delta t \neq 0$.

Observe, moreover, that since the quotient

$$\frac{\Delta t(128 - 32t - 16\Delta t)}{\Delta t}$$

is equal to $128 - 32t - 16\Delta t$ for all Δt except when $\Delta t = 0$, and since $128 - 32t - 16\Delta t$ closes in on $128 - 32t$ when Δt gets close to zero, then it must follow that the quotient also approaches $128 - 32t$, which we also write as

$$\lim_{\Delta t \to 0} \frac{\Delta t(128 - 32t - 16\Delta t)}{\Delta t} = 128 - 32t$$

Now what does this mean for our actual problem? As the time intervals get smaller and smaller, or as the elapsed time gets more and more like an instant of time, the average velocity gets more and more like $128 - 32t$. With this in mind, mathematicians and physicists have *defined* the instantaneous velocity at the time t, for the position function $128t - 16t^2$, to be

$$\lim_{\Delta t \to 0} \frac{\Delta t(128 - 32t - 16\Delta t)}{\Delta t}$$

which, as we have seen, is $128 - 32t$.

The process of definition is quite general. Suppose we have a position function for a particle, which we write as $d(t)$. Then in the time interval $(t,\ t + \Delta t)$, the particle will start at $d(t)$ and wind up at $d(t + \Delta t)$, so that its average velocity is

$$\frac{d(t + \Delta t) - d(t)}{\Delta t}$$

and its instantaneous velocity at time t is *defined to be*

$$\lim_{\Delta t \to 0} \frac{d(t + \Delta t) - d(t)}{\Delta t}$$

Now, let's pose another problem which appears to be quite unrelated to the problem of instantaneous velocity. Suppose we wish to define the line tangent to a particular curve at a given point. If the curve is a circle, there is no difficulty; a tangent line is a line which intersects the circle exactly once. But suppose we are dealing with the graph of $y = x^3 - 9x$. The curve will look like Figure 10.1.

A line we want to be a tangent line at some point near the top of the hump will look like the one we have drawn in Figure 10.2; but such a line will also intersect the curve in at least one other point. Clearly, we must look elsewhere for our definition of tangent line.

Recall that in order to completely determine a straight line we need two pieces of information. Either we must have two points through which the line passes, or we must know one point and something about the angle it makes with the x axis, usually the value of the tangent of that angle, called the *slope* of the line. We will wind up with the point-and-slope formulation in our definition, but use the two-

Figure 10.1

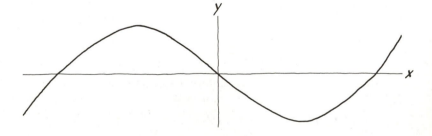

CHAPTER 10 / CALCULUS: A SAMPLER

Figure 10.2

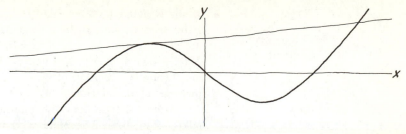

point method to arrive at it.

To fix our ideas, suppose we wish to define (and find) the line tangent to the graph of $y = x^3 - 9x$ at $(-2, 10)$. For the sake of clarity, we draw an enlarged portion of the graph near that point (Figure 10.3).

We consider first a secant line through $(-2, 10)$ and some nearby point which we have drawn. If we denote the difference in the x coordinate between the points of intersection by Δx, then the x coordinate of the second point is $(-2 + \Delta x)$. Then the y coordinate, is the value we obtain when we substitute $(-2 + \Delta x)$ into $x^3 - 9x$, namely $(-2 + \Delta x)^3 - 9(-2 + \Delta x)$. Then the slope[1] of this secant line is

$$\frac{[(-2 + \Delta x)^3 - 9(-2 + \Delta x)] - (10)}{-2 + \Delta x - (-2)}$$

The expression in square brackets is y_2, the 10 is y_1, x_1 is -2, and $-2 + \Delta x$ is x_2.

Doing the algebra (multiplying out, simplifying, and factoring) gives us that, for the given curve, the *slope of a secant line* through $(-2, 10)$ is

$$\frac{\Delta x \left(3 - 6\Delta x + (\Delta x)^2\right)}{\Delta x}$$

Figure 10.3

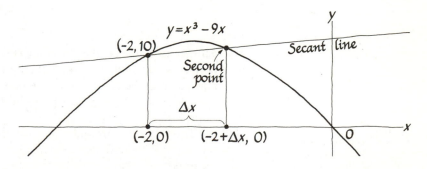

---

[1]Recall that the formula for the slope of the line between (x_1, y_1) and (x_2, y_2) is $(y_2 - y_1)/(x_2 - x_1)$.

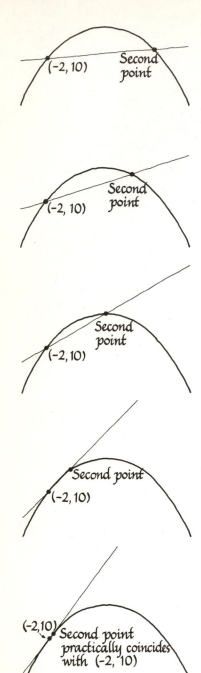

(-2, 10) Second point

(-2, 10) Second point

(-2, 10) Second point

Second point
(-2, 10)

(-2,10) Second point practically coincides with (-2, 10)

Figure 10.4

which is equal to $3 - 6\Delta x + (\Delta x)^2$, since $\Delta x \neq 0$. For example, if the second point is $(0,0)$, then Δx is 2 and the slope of the secant line is $3 - 6 \cdot 2 + 2^2 = -5$. Observe that we have found in $(\Delta x(3 - 6\Delta x + (\Delta x)^2))/\Delta x$ a general formula for the slope of any secant line through $(-2, 10)$, but we can replace it by $3 - 6\Delta x + (\Delta x)^2$ for purposes of calculation.

Observe that *at* $\Delta x = 0$, neither the algebra nor the geometry makes any sense. The algebra is meaningless because we have an expression involving division by zero. The geometry becomes nonsense since the second point needed to determine the line has disappeared. One single point cannot determine a line by itself without more information.

We can investigate these secant lines for values of Δx *near* zero. Geometrically, that means that the second point must be near to the first, $(-2, 10)$. Intuitively, we can see, in fact, that as the two points get closer together, the secant lines determined by them will approach more nearly the line which we feel should be the tangent line, as Figure 10.4 shows. However, as these secant lines get close to the tangent (that is, as Δx gets close to zero), the expression for the slope, $3 - 6\Delta x + (\Delta x)^2$, gets close to 3. To see this, $6\Delta x$ will be small if Δx is close to zero; $(\Delta x)^2$ will be even smaller since the product of two numbers close to zero is closer than either. (See Exercise 10.2.14.) But this means that

$$\frac{\Delta x(3 - 6\Delta x + (\Delta x)^2)}{\Delta x}$$

is also closing in on 3 as Δx nears zero, since the two expressions are equal everywhere except when $\Delta x = 0$. We write this as

$$\lim_{\Delta x \to 0} \frac{\Delta x(3 - 6\Delta x + (\Delta x)^2)}{\Delta x} = 3$$

This, in turn, means that the slopes of the secant lines approach 3 for nonzero values of Δx; and from the intuitive feeling of the secants approaching the tangent line, we now *define* the tangent to the graph of $y = x^3 - 9x$ at $(-2, 10)$ to be the line through $(-2, 10)$ with slope 3.

The processes are quite general. Suppose we take any point (x, y) on $y = x^3 - 9x$. Then a secant line through that point will have as its slope

$$\frac{[(x + \Delta x)^3 - 9(x + \Delta x)] - (x^3 - 9x)}{x + \Delta x - x}$$

where, as before, the expression in square brackets is y_2 of the slope formula, the matter in the round brackets represents y_1; $x + \Delta x$ is x_2 and x is x_1. This simplifies into

$$\frac{\Delta x(3x^2 - 9 + 3x(\Delta x) + (\Delta x)^2)}{\Delta x}$$

which is equal to $3x^2 - 9 + 3x\Delta x + (\Delta x^2)$ whenever $\Delta x \neq 0$, so that

$$\lim_{\Delta x \to 0} \frac{\Delta x(3x^2 - 9 + 3x\Delta x + (\Delta x^2))}{\Delta x} = 3x^2 - 9$$

What we have here is a general formula which gives us the slope of the tangent to $y = x^3 - 9x$ at any point (x, y) on its graph.

Thus, for example, at a different point, say $(1, -8)$, we have $x = 1$, so that the slope of the tangent is $3(1)^2 - 9 = 3 - 9 = -6$. If we put the tangent line in the form $y = mx + b$ (see Chapter 8), we can write the line as

$$y = -6x + b$$

since we already know that m, its slope, is -6. To find b, we use the fact that the line goes through $(1, -8)$, so that we must have that $y = -8$ if x is 1. This gives us

$$-8 = -6 \cdot 1 + b$$

or

$$b = -2$$

so that our line is

$$y = -6x - 2$$

We can generalize this entire procedure still further. Suppose we have a mathematical expression which we can write as $y = f(x)$ and are asked to find a general formula for the slope of the line tangent to its graph at (x, y). We start by noting that at $x + \Delta x$, the value of the y coordinate will be $f(x + \Delta x)$, so that the formula for the slope of any secant line through (x, y) is

$$\frac{f(x + \Delta x) - f(x)}{(x + \Delta x) - x} = \frac{f(x + \Delta x) - f(x)}{\Delta x}$$

and we *define* the tangent line to be the line through (x, y) whose slope is

$$\lim_{\Delta x \to 0} \frac{f(x + \Delta x) - f(x)}{\Delta x}$$

You will have noticed that this formula is exactly the same (except for notation) as that for instantaneous velocity. Now this is very interesting. Here we started at what appeared to be two separate and distinct problems, yet wound up with precisely the same solution for each. It turns out that there are any number of situations from reality which will lead us to this abstract formulation for their solutions (see, for example, Exercise 10.2.8).

Whenever mathematicians encounter several paths leading to the same abstraction, they consider the abstraction itself important enough to study, and we shall do exactly this. The expression

$$\lim_{\Delta x \to 0} \frac{f(x + \Delta x) - f(x)}{\Delta x}$$

is called *the derivative of f at x*; and traditionally there are several notations for it:

$$\frac{df}{dx}, \quad \frac{dy}{dx}, \quad D_x f, \quad D_x y, \quad f'(x)$$

are the chief ones in current usage.

EXERCISES 10.2

1. Given the position function $-16t^2$, find the average velocities between
   (a) $t = 0$ and $t = 1$
   (b) $t = 0$ and $t = 2$
   (c) $t = 0$ and $t = 4$
   (d) Compare your answer with the velocity at the halfway point in each time period. (Recall that the velocity at time t is $v = 32t$.)

2. Given the function $9t^3$, answer questions (a), (b), and (c) in Exercise 1.

3. Complete the average velocities for the position function $d = 128t - 16t^2$ for:
   (a) t between 0 and 8   (b) t between 1 and 7
   (c) t between 2 and 6   (d) t between 3 and 5
   (e) Is it reasonable to interpret your answers to (a), (b), (c), and (d) as indicating that the particle was motionless throughout the time period? Explain.

4. Carry out the algebra which was necessary as the first step for simplifying the average velocity in the interval $(t, t + \Delta t)$ for $d = 128t - 16t^2$.

5. Use the formula for average velocity $v_{av} = 128 - 32t - 16\Delta t$ to calculate the average velocities for:
   (a) t between 1 and 4   (b) t between 4 and 6
   (c) t between 0 and 8   (d) t between 0 and 4
   (e) t between 4 and 8

6. If Δt is negative, the interval of time becomes $(t + \Delta t, t)$ and the formula for average velocity remains the same.
   (a) Show that this is true for the position function $128t - 16t^2$.
   (b) Calculate the average velocity for t between 1 and 4, taking $t = 4$ and $\Delta t = -3$.

7. Suppose we have a position function $d(t) = 30 + 7t - t^2$.
   (a) Show that the average velocity in the time interval $(t, t + \Delta t)$ is $7 - 2t - \Delta t$.
   (b) Find the average velocity for t between 1 and 2; t between 1 and 3/2; t between 1 and 5/4; t between 1 and 17/16.
   (c) Find the instantaneous velocity at time t.
   (d) Interpret the results by assuming that the position formula gives the height above ground of a ball rolling up and down a large ramp where d measures the number of inches the ball is above the ground.
      1. Where is the ball at the start?
      2. How long does the ball rise?
      3. How high does it rise?
      4. What is the velocity at the start?
      5. What is the meaning of $d(t) = 0$?
      6. When does the ball hit the ground?
      7. What is its velocity when it hits the ground?

8. When we try to define *instantaneous acceleration* of a particle we run into the same difficulty as we did with instantaneous velocity. The solution is somewhat the same, particularly when the acceleration is not a constant, as for example, in your automobile. Suppose a particle has a velocity at time t given by $v(t) = 4t - t^2$.
   (a) Define the *average acceleration* in a time interval to be the velocity at the end minus the velocity at the beginning, divided by the length of the interval.
      1. Find the average acceleration for t between 0 and 3; between 1 and 4; between 1 1/2 and 6.

2. Find a general formula for the average velocity for the interval $(t, t + \Delta t)$.

(b) Interpret the results.
1. What is the meaning of $v(t) = 0$?
2. For what values of t is the particle going in a positive direction? a negative direction?
3. For what values of t is the particle being speeded up? retarded?
4. For what values of t is no force being applied to it?
5. Conceive of a physical model which would produce these results.

9. Carry out the computation to show that the slope of the secant line in the tangent problem of the text is $\dfrac{\Delta x(3 - 6\Delta x + (\Delta x)^2)}{\Delta x}$.

10. In the tangent problem of the text, if Δx is negative, then the second point is to the left of $(-2, 10)$. Show that the formula for the slope of a secant line remains unchanged.

11. Using the formula in the text or in Exercise 9, calculate the slope of the secant line for $\Delta x = 1, .1, -.01$, and $.001$.

12. Use the formula derived in the text to find the lines tangent to $y = x^3 - 9x$ at the following points.
(a) $(0, 0)$ (b) $(-1, 8)$ (c) $(3, 0)$
(d) $(-3, 1)$ (e) at $x = 4$ (f) at $x = \sqrt{3}$
(g) Look at your answer to (d). It cannot possibly be correct, since something is wrong with the question. What is it?

13. Show that the general formula for the slope of the line tangent to $y = x^2 - 6x + 3$ at (x, y) is $2x - 6$, and find the lines tangent to the graph at $x = -3$, $x = 0$, $x = 3$, and $x = 4$.

14. Compute $6\Delta x$ and $(\Delta x)^2$ for Δx equal to
(a) $.5$ (b) $.05$ (c) $.01$
(d) $.005$ (e) $.0001$ (f) $-.5$
(g) $-.01$ (h) $-.005$ (i) $-.000001$

10.3 The Limit

Before we can start our study of the derivative, we must, as usual, specify our undefined terms and indicate our axioms. Our principal undefined term is *limit* and what is meant by $\lim_{z \to a} f(z) = L$. The idea is this: we start with an algebraic expression f that depends on z in some way. Then, if the values of f close in on L as z gets close to a, we say that L *is the limit of f as z goes to a.*[1] That is, $\lim_{z \to a} f(z) = L$ means that we can make the values $f(z)$ as nearly equal to L as we please, by picking z sufficiently close to a.

(Observe that we are not interested in any way in what happens when z is *equal* to a. The expression f may not even be defined when $z = a$.)

Besides the examples we have already seen (with $z = \triangle t$ or $z = \triangle x$), suppose we consider

$$\lim_{x \to 2} \frac{x^2 - 6x + 8}{x - 2}$$

In this expression, the z of the definition is replaced by x, and the a by 2. Observe that the expression

$$\frac{x^2 - 6x + 8}{x - 2}$$

isn't defined at $x = 2$, but we *can* ask if it has a limit there. That means that we are interested in what happens to $(x^2 - 6x + 8)/(x - 2)$ when x gets arbitrarily close to 2 but without actually equalling 2.

To pursue this interest, we notice that the numerator can be factored into $(x - 2)(x - 4)$, so that, for $x \neq 2$,

$$\frac{x^2 - 6x + 8}{x - 2} = x - 4$$

Intuitively it is clear that as x nears 2, $x - 4$ closes in on -2, so that

$$\frac{x^2 - 6x + 8}{x - 2}$$

also closes in on -2. We write this as

$$\lim_{x \to 2} \frac{x^2 - 6x + 8}{x - 2} = -2$$

We can test this by trying a few examples. Thus, when x is 2.1 we get

---

[1]This notion of limit can actually be defined in more advanced work. However, the machinery of its definition is complicated, and making this machinery work in order to prove as theorems what we will assume below as axioms is very difficult and requires a deep knowledge of the real-number system.

$$\frac{x^2 - 6x + 8}{x - 2} = \frac{(2.1)^2 - 6(2.1) + 8}{2.1 - 2}$$

$$= \frac{4.41 - 12.6 + 8}{.1}$$

$$= \frac{-.19}{.1} = -1.9$$

Similar calculations show that when x is 2.01, the fraction is -1.99; and when x is 1.999, the fraction is -2.001. This illustrates an important principle which we list as:

AXIOM 10.1   *If $f(z) = g(z)$ for all z except possibly at $z = a$, and if $\lim_{z \to a} g(z) = L$, then $\lim_{z \to a} f(z) = L$ also.*

In our example above, the g of Axiom 10.1 is $x - 4$, and the f of the axiom is $(x^2 - 6x + 8)/(x - 2)$.

There are several other axioms which we used in that example.

AXIOM 10.2   *$\lim_{z \to a} z = a$.*

That is, as z gets close to a, z gets close to a. Another is

AXIOM 10.3   *If $f(z)$ is identically equal to some constant k, then $\lim_{z \to a} f(z) = k$.*

We abbreviate this by writing $\lim_{z \to a} k = k$. Furthermore,

AXIOM 10.4   *If $\lim_{z \to a} f(z) = L$ and $\lim_{z \to a} h(z) = M$, then $\lim_{z \to a} (f(z) + h(z)) = L + M$.*

Another way of writing Axiom 10.4 is

$$\lim_{z \to a} (f(z) + h(z)) = \lim_{z \to a} f(z) + \lim_{z \to a} h(z),$$

provided both limits on the right exist.

We show how we have already applied Axioms 10.2 through 10.4 to find that $\lim_{x \to 2} x - 4 = -2$. Observe that $x - 4 = x + (-4)$, also, by Axiom 10.2, $\lim_{x \to 2} x = 2$, and by Axiom 10.3, $\lim_{x \to 2} (-4) = -4$, and finally, by Axiom 10.4,

$$\lim_{x \to 2} (x - 4) = \lim_{x \to 2} x + \lim_{x \to 2} (-4) = 2 + (-4) = -2$$

It is now time to show that there are times when limits do not exist. Consider $\lim_{z \to 0} 1/z$. Let us actually calculate values for $1/z$ as z nears zero. Thus, for $z = 1/100$, $1/z = 100$; for $z = 1/1,000,000$, $1/z = 1,000,000$; while for $z = -1/100$, $1/z = -100$ and for $z = -1/1,000,000$, $1/z = -1,000,000$. So we see that, in this case, as z gets close to zero, the values of $1/z$ actually get farther and farther apart, so the limit cannot exist.

This illustrates yet another important point. The values of f must approach L when z gets close to *either side* of a, that is, when z nears a from values which are larger than a, as well as from those that are smaller. To see how this works, consider[1]

$$\lim_{z \to 0} \frac{\sqrt{z^2}}{z}$$

Now, $\sqrt{z^2}$ indicates the positive square root[2]; that is, $\sqrt{z^2}$ can never be negative. Thus, only for positive z and $z = 0$ is $\sqrt{z^2} = z$. For negative z, $\sqrt{z^2} = -z$. For example, $\sqrt{(-3)^2} = \sqrt{9} = 3 = -(-3)$. This gives us that

$$\frac{\sqrt{z^2}}{z} = 1 \qquad \text{for nonnegative } z,$$

and

$$\frac{\sqrt{z^2}}{z} = -1 \qquad \text{for negative } z.$$

Thus, as z goes to zero through positive values, $\sqrt{z^2}/z$ is 1, while $\sqrt{z^2}/z$ is -1 when z goes to zero through negative values. The result is that $\sqrt{z^2}/z$ does not close in on any single value as z nears zero, since, no matter how small an interval about zero we select, $\sqrt{z^2}/z$ is 1 if z is on one side of it and -1 if z is on the other. Thus, it cannot have a limit at zero.

Again, numerical examples may help. For $z = .01$, $\sqrt{(.01)^2} = .01$, and

$$\frac{\sqrt{z^2}}{z} = \frac{\sqrt{(.01)^2}}{.01} = \frac{.01}{.01} = 1$$

---

[1]Recall, $\sqrt{z^2} = |z|$, so we could rewrite this as $\lim_{z \to 0} |z|/z$.

[2]Recall also that if any mathematical symbol occurs with *no* sign, the plus sign is always assumed.

while, for $z = -.001$, $\sqrt{(-.001)^2} = .001$, so that

$$\frac{\sqrt{z^2}}{z} = \frac{.001}{-.001} = -1$$

AXIOM 10.5 *If $\lim_{z \to a} f(z) = L$ and $\lim_{z \to a} h(z) = M$, then $\lim_{z \to a} (f(z)h(z)) = LM$.*

Another way of writing Axiom 10.5 is

$$\lim_{z \to a} (f(z)h(z)) = (\lim_{z \to a} f(z))(\lim_{z \to a} (h(z))$$

whenever both limits on the right exist. Notice that this has the immediate consequence that $\lim_{z \to a} z^2 = a^2$ (by Axioms 10.5 and 10.2), from which we see that

$$\lim_{z \to a} z^3 = (\lim_{z \to a} z^2) \cdot (\lim_{z \to a} z) = a^2 \cdot a = a^3$$

and so on. Thus we can conclude that, for any positive integer n, $\lim_{z \to a} z^n = a^n$. Also, we observe that, if k is any constant, and if $\lim_{z \to a} f(z) = L$, then

$$\lim_{z \to a} kf(z) = kL \qquad \text{or} \qquad \lim_{z \to a} kf(z) = k \lim_{z \to a} f(z).$$

Thus, we have a result important enough to record as a theorem:

THEOREM 10.1 $Lim_{z \to a} kz^n = ka^n$, *where n is any positive integer and k is any constant.*

Finally, we have

AXIOM 10.6   *If $\lim_{z \to a} f(z) = L$ and $\lim_{z \to a} h(z) = M$, then*

$$\lim_{x \to a} \left(\frac{f(z)}{h(z)}\right) = \frac{L}{M} \qquad \text{whenever } M \text{ is not zero.}$$

As with the other axioms, this too can be rewritten, this one as

$$\lim_{z \to a} \left(\frac{f(z)}{h(z)}\right) = \frac{\lim_{z \to a} f(z)}{\lim_{z \to a} h(z)}$$

provided that both limits exist and the denominator limit is not zero.

Axiom 10.6 means that we can now extend Theorem 10.1 to *all* integers n if a is not zero, since

$$\lim_{z \to a}\left(\frac{1}{z^n}\right) = \frac{1}{a^n}$$

EXERCISES 10.3

1. Find the values of the fraction $(x^2 - 6x + 8)/(x - 2)$ when
   (a) $x = 1.9$ (b) $x = 1.99$
   (c) $x = 2.001$ (d) 2.000000001

2. Find the values of the fraction $(x^2 + 99x - 100)/(x - 1)$ for
   (a) $x = 2$ (b) $x = 1.1$ (c) $x = 1.01$
   (d) $x = 0.9$ (e) $x = .99$ (f) $x = .999$

   (g) What do your answers suggest about $\lim\limits_{x \to 1} \dfrac{x^2 + 100x - 100}{x - 1}$?

3. Go to a good table for e^x (your library has one) and find the values for $(e^x - 1)/x$ for
   (a) $x = 1$ (b) $x = .5$ (c) $x = .1$
   (d) $x = .05$ (e) $x = .02$ (f) $x = .01$
   (g) $x = -1$ (h) $x = -.5$ (i) $x = -.1$
   (j) $x = -.05$ (k) $x = -.02$ (l) $x = -.01$
   (m) What do your results suggest about $\lim_{x \to 0} (e^x - 1)/x$?

4. Show that $\lim_{z \to a} kf(z) = k \lim_{z \to a} f(z)$, indicating clearly what axioms are being used.

5. Show that $\lim_{x \to a} (f(x) - g(x)) = \lim_{x \to a} f(x) - \lim_{x \to a} g(x)$.

6. Show that $\lim_{z \to 0} 1/z^2$ does not exist.

7. Find the following limits, where they exist. If they do not exist, see if you can explain why.
   (a) $\lim_{x \to 3} x - 2$ (b) $\lim_{z \to 1} 3z^2 + 6$

   (c) $\lim_{z \to -1} \dfrac{z^2 - 1}{z + 1}$ (d) $\lim_{z \to 1} \dfrac{3z^3 + 1 - 4z}{z + 1}$

   (e) $\lim_{z \to 1} \dfrac{3z^2 + 1}{z - 1}$ °(f) $\lim_{z \to 1} \dfrac{3z^3 + 1 - 4z}{z - 1}$

8. Prove Theorem 10.1, or its extension, in detail for
   (a) $n = 5$
   (b) $n = -3$

9. If you know that $\lim_{z \to 0} (\sin z)/z = 1$, find
   (a) $\lim_{z \to 0} \dfrac{z}{\sin z}$ (b) $\lim_{z \to 0} \dfrac{3 \sin z}{2z}$
   °(c) $\lim_{z \to 0} \dfrac{\sin (3z)}{2z}$

10. We say that f has the limit L, *from the right* at a if the values of $f(z)$ close in on L when z gets close to a through values larger than a, and *from the left* if f nears L as z gets close to a through values smaller than a. Thus $\sqrt{z^2} / z$ has a righthand limit of 1 and a lefthand limit of -1.
    (a) Find the righthand limits at zero (when they exist) of $z/\sqrt{z^2}$, $(1 - (1 - \sqrt{z^2}))/z$, $1/z$, $1/z^2$, $\sqrt{z}$.
    (b) Find the lefthand limits at zero (when they exist) of $z/\sqrt{z^2}$, $1/z$, $1/z^2$, $\sqrt{z}$, $\sqrt{-z}$.
    °(c) Is it true that the limit of f at a equals L when and only when the left and right limits of f both exist and are both L?

10.4 Differentiating

As we have already seen, the collection of symbols

$$\lim_{\Delta x \to 0} \frac{f(x + \Delta x) - f(x)}{\Delta x}$$

which we called *the derivative of f at x*, and which can be written $D_x f$, whenever the limit exists, is a mathematical abstraction representing several situations from reality. If the expression f is at all complicated, it is tedious in the extreme to compute its numerator. Furthermore, we cannot directly apply Axioms 10.2 through 10.6 to the limit, since the denominator limit is zero. However, there are certain formulas which we can derive to help us in our calculations, so that it will not be necessary to always return to the definition. This process is called *differentiation*.

   We begin by showing that $D_x x = 1$; that is, the derivative of the expression $f(x) = x$ is 1. Since $f(x) = x$, then $f(x + \Delta x) = x + \Delta x$, so that

$$\frac{f(x + \Delta x) - f(x)}{\Delta x} = \frac{x + \Delta x - x}{\Delta x} = \frac{\Delta x}{\Delta x} = 1$$

whenever $\Delta x \neq 0$. Thus, by Axioms 10.3 and 10.1, we have

$$\lim_{\Delta x \to 0} \frac{x + \Delta x - x}{\Delta x} = 1$$

which is the same as saying $D_x x = 1$. Next we can show that:

THEOREM 10.2 $D_x(f(x) + g(x)) = D_x f(x) + D_x g(x)$, or, in words, the derivative of the sum is the sum of the derivative.

Proof[1] $$D_x(f(x) + g(x)) = \lim_{\Delta x \to 0} \frac{[f(x + \Delta x) + g(x + \Delta x)] - [f(x) + g(x)]}{\Delta x}$$

which becomes, after rearranging the numerator,

$$\lim_{\Delta x \to 0} \frac{[f(x + \Delta x) - f(x)] + [g(x + \Delta x) - g(x)]}{\Delta x}$$

This turns into

$$\lim_{\Delta x \to 0} \frac{[f(x + \Delta x) - f(x)]}{\Delta x} + \frac{[g(x + \Delta x) - g(x)]}{\Delta x}$$

since

$$\frac{A + B}{C} = \frac{A}{C} + \frac{B}{C}$$

This in turn, becomes

$$\lim_{\Delta x \to 0} \frac{f(x + \Delta x) - f(x)}{\Delta x} + \lim_{\Delta x \to 0} \frac{g(x + \Delta x) - g(x)}{\Delta x}$$

by Axiom 10.4. Thus $D_x(f(x) + g(x)) = D_x f(x) + D_x g(x)$, by the definition of derivative. $\square$

Notice, incidentally, that only the last step involves limit theorems; the earlier ones are all algebraic in nature.

Theorem 10.2 allows us to assert that $D_x(2x) = 2$, since $D_x(2x) = D_x(x + x) = D_x x + D_x x = 1 + 1 = 2$. But we have another way of proving the same thing.

---

[1]The proofs are included for the curious. Understanding them is not necessary for understanding the rest of the chapter.

THEOREM 10.3 $D_x kf(x) = kD_x f(x)$ *for any constant k. This assumes that $D_x f$ exists.*

Proof   We have

$$D_x kf(x) = \lim_{\Delta x \to 0} \frac{kf(x + \Delta x) - kf(x)}{\Delta x}$$

which, after factoring in the numerator, becomes

$$\lim_{\Delta x \to 0} \frac{k(f(x + \Delta x) - f(x))}{\Delta x}$$

so that, by Axioms 10.5 and 10.3, we get

$$k \lim_{\Delta x \to 0} \frac{f(x + \Delta x) - f(x)}{\Delta x} = kD_x f(x)$$

by the definition of derivative. □

This gives us not only that $D_x(2x) = 2D_x(x) = 2$, but more generally, that $D_x kx = kD_x x = k \cdot 1 = k$.

It would be pleasant if, analogous to Theorem 10.2, $D_x(f(x)g(x)) = (D_x f(x))(D_x g(x))$. Suppose we try it on $D_x(x^2)$:

$$D_x(x^2) = (D_x x)(D_x x) = 1 \cdot 1 = 1$$

But we can compute $D_x(x^2)$ directly from the definition:

$$D_x(x^2) = \lim_{\Delta x \to 0} \frac{(x + \Delta x)^2 - x^2}{\Delta x}$$

$$= \lim_{\Delta x \to 0} \frac{\cancel{x^2} + 2x\Delta x + (\Delta x)^2 - \cancel{x^2}}{\Delta x}$$

$$= \lim_{\Delta x \to 0} \frac{\cancel{\Delta x}(2x + \Delta x)}{\cancel{\Delta x}}$$

$$= \lim_{\Delta x \to 0} 2x + \Delta x = \lim_{\Delta x \to 0} 2x + \lim_{\Delta x \to 0} \Delta x$$

$$= 2x + 0 = 2x$$

Now $2x$ is not the same as 1, so we must conclude that, as a general rule, $D_x(f(x)g(x)) = (D_x f(x))(D_x g(x))$ must be false. It would be hard to guess at the correct formula, which we give in the next theorem.

THEOREM 10.4 $D_x(f(x)g(x)) = f(x)D_x g(x) + g(x)D_x f(x).$

For example, $D_x x^2 = D_x(x \cdot x) = x D_x x + x D_x x = x \cdot 1 + x \cdot 1 = 2x$.

We shall not prove this theorem, but shall use it to draw some important conclusions.

COROLLARY 1 $D_x k = 0$.

That is, the derivative of a constant is always zero. This makes perfectly good intuitive sense when we recall that the derivative represents the measurment of, say, velocity. If a particle is *constantly* in the same place, its velocity is zero.

Proof   By Theorem 10.4,

$$D_x kx = x D_x k + k D_x x = x D_x k + k$$

and by Theorem 10.3, we also have

$$D_x kx = k D_x x = k$$

Hence, setting things which are equal to the same thing equal to each other, and canceling,

$$k = x D_x k + k$$

or

$$x D_x k = 0$$

Since it is true for any x, it must be true when $x = 1$, which gives $D_x k = 0$ as promised.

COROLLARY 2 $D_x x^2 = 2x$, $D_x x^3 = 3x^2$, $D_x x^4 = 4x^3$, etc., *and, in general,*

$$D_x(x^n) = n x^{n-1}$$

for integer n.

Proof   We have already shown that $D_x x^2 = 2x$. Using Theorem 10.4, $D_x x^3 = D_x(x \cdot x^2) = x D_x x^2 + x^2 D_x x = x \cdot 2x + x^2 \cdot 1 = 2x^2 + x^2 = 3x^2$, and $D_x x^4 = D_x(x \cdot x^3) = x D_x x^3 + x^3 D_x x = x \cdot 3x^2 + x^3 \cdot 1 = 3x^3 + x^3 = 4x^3$. This process is easily continued, so that we can show that $D_x x^5 = 5x^4$, $D_x x^6 = 6x^5$, and so on. In fact, the rule is quite general, i.e., $D_x x^n = n x^{n-1}$ for any positive integer n. We can then extend it to negative integer n by observing that if n is a negative integer, $-n$ is positive and $x^n \cdot x^{-n} = 1$. Then, by differentiating both sides, $D_x(x^n \cdot x^{-n}) = D_x(1) = 0$. But also, $D_x(x^n \cdot x^{-n}) = x^n Dx(x^{-n}) + x^{-n}D_x(x^n)$, so that $x^n D_x(x^{-n}) + x^{-n}D_x(x^n) = 0$. But $D_x(x^{-n}) = -nx^{-n-1}$ as $-n$ is positive. Substituting gives

$$x^n(-nx^{-n-1}) + x^{-n}D_x(x^n) = 0$$

or

$$-nx^{-1} + x^{-n}D_x(x^n) = 0$$

or

$$x^{-n}D_x(x^n) = nx^{-1}$$

or

$$D_x(x^n) = (nx^{-1})x^n = nx^{n-1}$$

EXERCISES 10.4

1. Supply the reasons for each step in the text derivation of $D_x x^2 = 2x$ from the definition of $D_x f(x)$.

°2. Show that $D_x(f(x) - g(x)) = D_x f(x) - D_x g(x)$.

3. Prove the formulas in the text for $D_x(x^5)$ and $D_x(x^6)$.

4. Is it true that $D_x(1/x) = -1/x^2$? $D_x(1/x^2) = -2/x^3$? Why?

5. Find $D_x(\sqrt{x})$ by noting that $x = \sqrt{x} \cdot \sqrt{x}$ if $x \geq 0$, and differentiating both sides.

6. Show that $D_x(fgh) = fg(D_x h) + fh\,(D_x g) + hg(D_x f)$. (*Hint:* Use the associative law and Theorem 10.4.)

°7. (a) Find the derivative of $\sqrt[3]{x}$. (*Hint:* Follow the pattern of Exercises 5 and 6.)
    (b) Find the derivative of $1/\sqrt{x}$.
    (c) Find the derivative of $\sqrt[4]{x}$. (*Hint:* $\sqrt[4]{x}\sqrt[4]{x} = \sqrt{x}$.)
    (d) Find the derivative of $1/\sqrt[3]{x}$.
    (e) Find the derivative of $\sqrt[3]{x^2}$.
    (f) What general conclusion do your answers in Exercises 3, 4, 5, and 7, parts (a) through (e) indicate?

10.5 The Derivative

We now have at our disposal several formulas, so that we can differentiate expressions of the form $6x^4 - x^2 + \frac{1}{4}x - 3$. We do this, once, in complete detail:

$$\begin{aligned}
D_x(6x^4 - x^2 + \tfrac{1}{4}x - 3) &= D_x(6x^4) - D_x(x^2) + D_x(\tfrac{1}{4}x) - D_x(3) \\
&= 6D_x(x^4) - D_x(x^2) + \tfrac{1}{4}D_x(x) - D_x(3) \\
&= 6 \cdot 4x^3 - 2x^1 + \tfrac{1}{4} \cdot 1 - 0 \\
&= 24x^3 - 2x + \tfrac{1}{4}
\end{aligned}$$

If we tried doing this from the definition, the work would cover several square feet of paper.

We can proceed now to put the derivative to work for us. Suppose, to fix our attention on a particular problem, that we have a particle whose distance above ground is given as a function of t, by $d(t) = -16t^2 + 240t$. Taking the derivative of d, $D_t d(t)$, we get

$$-16(2t) + 240 \cdot 1 = -32t + 240$$

In this case we know $D_t(d(t))$ represents the velocity at time t, measured in feet per second. Thus, when $t = 0$ (at the start of the experiment), the velocity is 240 feet per second. That is, the force applied to the particle was enough to send it upwards at 240 feet per second. The experiment is over, some 15 seconds later, when the particle hits the ground; then the velocity is $-32(15) + 240 = -480 + 240 = -240$ feet per second. We also notice that the velocity is zero at $t = 7.5$ at which time it reaches its maximum altitude of 900 feet, and that for t between 0 and 7.5 the derivative, and so the velocity, is positive, indicating that the particle is rising. For t larger than 7.5, the negative sign of the derivative shows that the particle is falling.

Next, suppose we consider what happens when we interpret the derivative as the slope of the tangent line. Again, suppose we look at a specific problem: the graph of $y = 2x^3 - 3x^2 - 36x + 20$. The derivative $D_x y$ is $6x^2 - 6x - 36$, and it is easier to attack if we write it in factored form:

$$6(x^2 - x + 6) = 6(x - 3)(x + 2)$$

We see at once that the derivative is zero when $x = 3$ and -2. We also note that, for x less than -2, the derivative is positive (both factors are negative), the derivative is negative for x between -2 and 3 (only one factor is negative), and is positive for all x larger than 3 (both factors are positive). So what?

Since the derivative represents the slope of the tangent line, if it is positive, then the line must slant this way: /. As the curve locally tends to follow along in the same general direction as its tangent, we can see that as x is increasing, y must also increase. To put it differ-

Figure 10.5

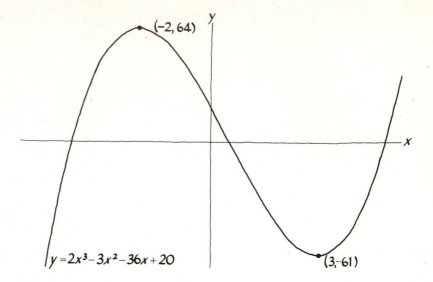

$y = 2x^3 - 3x^2 - 36x + 20$

(−2, 64)

(3, −61)

ently, as the x coordinate moves from left to right, the y coordinate will generally be rising near points where the derivative is positive.

Furthermore, if the derivative is negative, then the tangent line must slope this way: \. This means that as x increases, the y coordinate must *decrease*. In other words, a negative derivative means that, as x moves from left to right, the y coordinate will have a downward drift.

What then happens when the derivative is zero? Here we have a horizontal tangent line, since a line with zero slope is horizontal. This means that the curve momentarily levels off at any point where the derivative vanishes. In our case this happens at $x = 3$ and -2. Let's examine these points with special care.

When $x = -2$, $y = 64$, and, as we have seen, the tangent is horizontal there. But, to the left of $(-2, 64)$ the slope is positive, and to the right of it, the slope is negative, so the graph is rising to this point, and then is falling away, giving it approximately a dome shape, like this: ⌒ , with the top of the dome at $(-2, 64)$. In other words, *locally*, the point $(-2, 64)$ is the highest point of the graph, and, still speaking only locally, the y coordinate, 64, is the *maximum* y value.

When $x = 3$, $y = -61$. Here the slope is negative to the left and positive to the right, so that the graph falls to the point $(3, -61)$, and then starts to rise, giving it a cup-like shape like this: ⌣ with $(3, -61)$ at its bottom. This means, locally speaking again, that $(3, -61)$ is the lowest or *minimum* point, and that, for nearby points, at least, the y value will not drop below -61.

If we put it all together: graph rising to $(-2, 64)$, falling between $(-2, 64)$ and $(3, -61)$, and then rising again, we have a picture which looks like Figure 10.5 and it is easy to visualize what we mean by local maximum and local minimum. (In Figure 10.5, we have been forced to use a smaller scale vertically than horizontally.)

We can draw some general conclusions from this discussion. For any graph, a positive derivative means a rising graph with an increasing y coordinate (as x increases), while a negative derivative means a decreasing y coordinate. A zero derivative indicates a point where y neither increases nor decreases.

Notice that any local maximum or minimum can occur only at points with horizontal tangents, but a horizontal tangent does not necessarily mean either a maximum or a minimum. Consider the graph of $y = x^3$. The derivative is $3x^2$, so there is a horizontal tangent at $(0, 0)$. But observe that the derivative is positive on both sides of it, and what happens to the graph is that it rises to $(0, 0)$, "rests" there, and then continues to rise. It looks like Figure 10.6.

The rule is quite general. *Local minimums or maximums can occur only when the derivative is zero and also changes sign. If the sign remains the same, there is neither maximum nor minimum.*

This has certain practical applications. For example, suppose I have 400 feet of fencing. I wish to fence in an area along a straight river bank where no fencing is needed, as in Figure 10.7. How do I arrange the fencing in order to enclose the largest possible rectangular area? We denote the area we wish to enclose by $A = xy$ in which case, from the diagram, we must have $2x + y = 400$ or $y = 400 - 2x$. Substituting for $y = 400 - 2x$ in A we have

$$A = x(400 - 2x) = 400x - 2x^2$$

In order to determine the local maximum, if any, we find that $D_x A = 400 - 4x$, which is zero only when $x = 100$. We also see that to the left of 100, $D_x A$ is positive, and to the right it is negative, giving us a local picture like this: ⌒ , a maximum. Hence the solution to the fencing problem is to make the sides which are perpendicular to the river, x, 100 feet, and the remaining side 200 feet; the maximum area is 20,000 square feet.

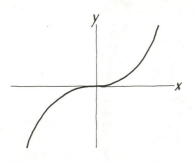

Figure 10.6

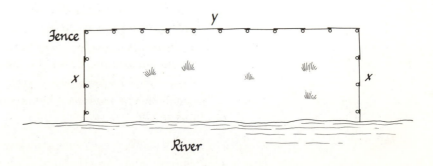

Figure 10.7

1. Justify each step in the text's differentiation of $6x^4 - x^2 + \frac{1}{4}x - 3$.

2. Suppose a ball is thrown upwards from the edge of a 656-foot-high cliff and lands at its foot, so that the distance d above the ground is $d(t) = -16t^2 + 640t + 656$, meaningful for t between zero and 41.
   (a) Verify that the conditions of the problem indicate that the expression for d makes no sense for negative t nor for t greater than 41.
   (b) How fast is the ball thrown upward?
   (c) What is its velocity when it hits the ground?
   (d) Where is its velocity zero? What does this mean in terms of the direction it is going?
   (e) What is the highest point reached?
   (f) This entire problem has been constructed with the origin at the foot of the cliff. Redo the problem with the orgin at the top of the cliff. How do you account for the different answers you get? Interpret the difference in terms of Chapter 1.

3. Consider the graph of $y = 3x^2 - x^3 - 2$.
   (a) Find its derivative; write it in factored form.
   (b) Determine the values of x for which the derivative is zero.
   (c) Determine where the derivative is positive, where negative?
   (d) Find any local maximum or minimum points.
   (e) Sketch the graph.

4. Analyze the graphs of the following as was done with $y = 2x^3 - 3x^2 - 36x + 20$ in the text, and graph them all on the same paper.
   (a) $y = 2x^3 - 3x^2 - 36x + 10$
   (b) $y = 2x^3 - 3x^2 - 36x - 3$
   (c) $y = 2x^3 - 3x^2 - 36x - 8$
   (d) $y = 2x^3 - 3x^2 - 36x + \sqrt{2}$
   What general conclusion can be drawn from the graphs of (a) through (d)?

5. The sum of two numbers must be 174; find the numbers so that their product is a maximum.

6. The product of two numbers is 144; find the numbers so that their sum is a maximum. (*Hint*: You may have to use Exercise 10.4.4)

7. I have 1000 feet of fencing. What is the largest rectangular area which can be enclosed? What is its shape? Can you generalize this result?

8. Find two numbers whose sum is 360 and which have the property that the product of the square of one times the other is a maximum.

9. Find the dimensions of the closed box of largest volume which can be constructed with 216 square feet of material for its sides if one end must be a square. What is this maximum volume? (Ignore waste, overlapping, etc.)

10. A closed box with one end square must contain 128 cubic feet. If the square ends cost $2 per square foot to make and the other sides cost $1 per square foot, what are the dimensions of the least expensive box? What is its cost? (*Hint*: You may have to use Exercise 10.4.4.)

11. Find the tangent lines to the graph of $y = 6x - 3$ at the following points.
    (a) $(0, -3)$ (b) $(2, 9)$
    (c) $(-3, -2)$ *(d) $(x_0, 6x_0 - 3)$
    (e) What general conclusion can you draw from your answers in (a) through (d)? Can you extend the result even further?

*10.6 Not Every Graph Has a Tangent at Every Point

As we have gone through the earlier sections of this chapter, we may be left with the impression that every graph has a tangent line at every point, or, which amounts to almost the same thing, that

$$\lim_{\Delta x \to 0} \left\{ \frac{f(x + \Delta x) - f(x)}{\Delta x} \right\}$$

will always exist for each x. To show that this is not so, consider the graph of $f(x) = \sqrt[3]{x^2} = x^{2/3}$. At $x = 0$, we have $f(x + \Delta x) = f(0 + \Delta x) = (0 + \Delta x)^{2/3} = (\Delta x)^{2/3}$, while $f(x) = f(0) = (0)^{2/3} = 0$. Thus, the quotient in braces above becomes

$$\frac{(\Delta x)^{2/3} - 0}{\Delta x} = \frac{(\Delta x)^{2/3}}{\Delta x} = \frac{1}{(\Delta x)^{1/3}}$$

What happens to this as Δx nears zero? As Δx goes to zero through positive values, $1/(\Delta x)^{1/3}$ gets very large; when Δx goes to zero through negative values, $1/(\Delta x)^{1/3}$ gets smaller and smaller so there can be no limit. (Recall in this context that $-1,000,000,000$ is smaller than $-1,000$.)

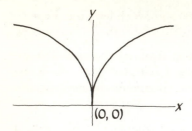

Figure 10.8

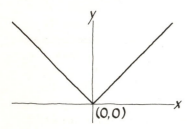

Figure 10.9

This is an example of a relatively rare occurrence; even though the derivative fails to exist at (0, 0) for this graph, there is a tangent line, a vertical tangent. Recall that a vertical line has no slope, so that this represents no exception to the general theory. The graph looks like Figure 10.8.

Generally, if there is no derivative, there is no tangent line. Consider the graph of $f(x) = \sqrt{x^2}$. (Recall from Section 10.3 that $\sqrt{x^2} = x$ if x is not negative and $\sqrt{x^2} = -x$ if x is negative.) At zero, $f(x + \Delta x) = f(0 + \Delta x) = \sqrt{(0 + \Delta x)^2} = \sqrt{(\Delta x)^2}$, while $f(x) = f(0) = \sqrt{0^2} = 0$. Thus the quotient defining the derivative is $\sqrt{(\Delta x)^2}/\Delta x$ and, as we saw in Section 10.3, this is 1 for positive Δx, -1 for negative Δx, and therefore has no limit as Δx approaches zero.

The graph of $\sqrt{x^2}$ looks like Figure 10.9 near zero; it has a *corner point* there.

Intuitively we see that there is no single line which would be a candidate for the tangent line; so we say, in light of the failure of the limit to exist, that no tangent line exists there. This is generally true: no tangent line at a corner point. One of the astonishing accomplishments of nineteenth-century mathematics was the construction of a curve with no tangent lines anywhere. Every point is a corner point!

10.7 Antidifferentiation

When we look at some of the simple operations of mathematics such as adding, multiplying, or squaring, we see that there are inverse operations which essentially *undo* whatever the original operations did. This process is called inverting, and in these cases the inverses are subtracting, dividing, and the taking of square roots. In calculus, too, there is an inverse to differentiation, called *antidifferentiation* or *integration*.

We say that $F(x)$ *is an antiderivative of* $f(x)$ if $D_x F(x) = f(x)$. That is, in this case we know the derivative and are trying to recover the original function that gave rise to it in the first place. These problems have enormous importance today, particularly in physics, and we will sample a few later on.

When we actually come to the process of antidifferentiation, we immediately encounter a difficulty of a new kind, one which we have not seen before. Suppose $F(x)$ is an antiderivative of $f(x)$. That is, $D_x F(x) = f(x)$. But then, so is $F(x) + 5$, since $D_x(F(x) + 5) = D_x F(x) + D_x 5 = D_x F(x) + 0 = D_x F(x) = f(x)$. Also, so is $F(x) - 3$ and $F(x) + \sqrt{3}$ and $F(x) - \sqrt[17]{\pi}$. In fact, so is $F(x) + C$, no matter *what constant value* C assumes. That is, there exists a whole family of solutions to the problem "find the antiderivative of $f(x)$." We will write the entire family as $F(x) + C$ and call it the *general antiderivative* of $f(x)$, and C the *constant of integration*.

Awkward as this constant may seem, antidifferentiation depends upon it to give us a useful abstraction of reality. For example, suppose we know that the upward velocity of a certain particle is $-32t + 640$. The general antiderivative is $-16t^2 + 640t + C$ (check this by taking its derivative), but this really does not give us a very complete picture of the position function corresponding to that velocity. In order to get a finished answer (in this case, the expression $d(t)$, the distance above the ground at time t), we need one further piece of information. To see this, suppose the particle started from the ground. Then, at time $t = 0$, $d = 0$, so that, by substituting in $d(t) = -16t^2 + 640t + C$, we have that $0 = d(0) = -16 \cdot 0^2 + 640 \cdot 0 + C = 0 + 0 + C$, or that $C = 0$. On the other hand, if it started from a 120-foot cliff, so that at $t = 0$, $d = 120$, then $120 = d(0) = 0 + 0 + C$ or $C = 120$. And, in fact, we can see that, in this case, C represents the height above ground of the particle when the experiment starts.

We might have to arrive at this indirectly. For example, suppose that we knew that the particle was at 4000 feet 10 seconds after it started. Then $d(10) = 4000$. That is, $4000 = -16 \cdot 100 + 640 \cdot 10 + C$, or $4000 = -1600 + 6400 + C$, or $4000 = 4800 + C$, or $C = -800$; we started from an 800-foot-hole in the ground. This discussion is the mathematical reflection of the physical fact that, in this problem, velocity depends *only* upon the initial velocity and the time in the air, not that height above the ground where the experiment started.

For the simple expressions we are dealing with, the rules of antidifferentiation are not difficult and are easily verified by differentiation.

The general antiderivative of kx^n is

$$\frac{kx^{n+1}}{n+1} + C$$

Observe that this breaks down for $n = -1$; the difficulty is that, to deal with the antiderivative of $1/x$, we would have to introduce a whole new family of functions. However, for any other value of n the rule holds. Thus the general antiderivative of $6x^7$ is

$$\frac{6x^{7+1}}{7+1} + C = \frac{6x^8}{8} + C = \frac{3x^8}{4} + C$$

Notice that the rule holds for $n = 0$, since the antiderivative of a constant k is $kx + C$.

If an antiderivative of f(x) is F(x) and if an antiderivative of g(x) is G(x), then the general antiderivative of f(x) + g(x) is F(x) + G(x) + C.

This means that the antiderivative of $x^3 - 6x^2 + 10x - 2$ is

$$\frac{x^4}{4} - 6\frac{x^3}{3} + 10\frac{x^2}{2} - 2x + C = \frac{x^4}{4} - 2x^3 + 5x^2 - 2x + C$$

which is easily verified by taking the derivative of the last line.

When it comes to taking the antiderivative of a product, there is no nice rule like the one we had for differentiation. We have no other recourse but to multiply out the product and then antidifferentiate. Similarly, quotients must be divided first.

Often the symbol "$\int \ldots \, dx$" is used for antidifferentiation. Thus, $F(x) + C = \int f(x) \, dx$ is read, "$F(x)$ is the general antiderivative of $f(x)$" or $G(x) = \int g(x) \, dx$ is read, "$G(x)$ is *an* antiderivative of $g(x)$." Using this notation we can write the antidifferentiation rules as

$$\int kx^n dx = \frac{kx^{n-1}}{n+1} + C$$

and

$$\int (f(x) + g(x)) \, dx = \int f(x) \, dx + \int g(x) \, dx$$
$$= F(x) + G(x) + C$$

If we interpret the derivative as the slope of the tangent line, then an antiderivative is an expression which gives rise to the original derivative in the first place. The general antiderivative will represent a whole family of curves, one for each specific antiderivative, all with the same shape, and all lying one above the other.

For example, consider $3x^2 + 6x$, which has as its general antiderivative $x^3 + 3x^2 + C$. In order to specify which particular curve of the family we want, we need to know one, but only one, point on the curve. For example, the curve through $(1, 2)$ means that $y = 2$ when $x = 1$; substituting that information into $y = x^3 + 3x^2 + C$, we get $2 = 1 + 3 + C$ or $C = -2$, so that the single curve of the family $x^3 + 3x^2 + C$ which passes through $(1, 2)$ is $x^3 + 3x^2 - 2$.

Notice that all curves of the family share those attributes described by the derivative. Thus all have horizontal tangents at $x = 0$ and $x = -2$, although the y coordinates will vary from curve to curve. Also, all will be increasing to $x = -2$, since the derivatives are positive, all will be decreasing for x between -2 and zero, and all will be increasing for positive x. We sketch a few of the curves in Figure 10.10.

Figure 10.10

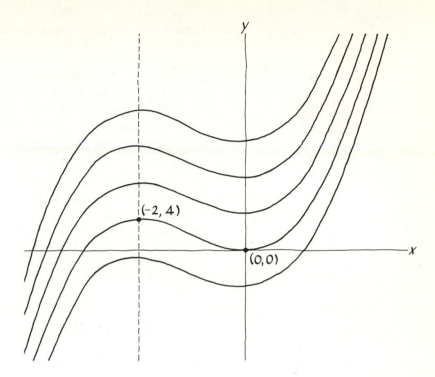

EXERCISES 10.7

1. (a) Show that $-16t^2 + 640t + C$ is the general antiderivative of $-32t + 640$.
   (b) Find the constant C if:
      1. $d(0) = 3$
      2. $d(1) = 4$
      3. $d(40) = 0$
   (c) Interpret the results of 1, 2, and 3 above if $-32t + 640$ is considered to be an expression for the velocity of a ball propelled upwards.

2. The vertical velocity of a ball rolling down a 1200-foot-high hill is $-2t - 10$ feet per second t seconds after it started.
   (a) Was the ball given a push to get it started?
   (b) Find the vertical distance above the ground after t seconds.
   (c) How long does it roll before it hits bottom?
   (d) How fast is it going downward when it hits bottom?

3. Consider the family of curves in the text, $x^3 + 3x^2 + C$.
   (a) Find C and sketch the curves which go through
       1. $(0, 0)$ 2. $(1, 0)$ 3. $(0, 1)$
       4. $(-2, 4)$ 5. $(0, -4)$
   (b) Is it true that every point in the plane has one curve of the family passing through it? Show that it is so or find a point with no such curve.

4. Consider the family of curves which have $4x(x^2 - 1)$ as their derivative.
   (a) Find the general expression for the family.
   (b) Find where the horizontal tangents are.
   (c) Find where each is increasing and each decreasing.
   (d) Find the particular curves through
       1. $(0, 0)$ 2. $(3, 0)$ 3. $(-3, 0)$
       4. $(2, 1)$ 5. $(2, 2)$

10.8 Further Applications of the Antiderivative

Suppose we wish to find the area under a curve, above the x axis and between two vertical lines. If the curve is a horizontal line, the area is a rectangle, and we *define* the area to be bh (base times height). If the area is triangular, it is not difficult to show that the area is $\frac{1}{2} bh$, and if the area is a region enclosed by a series of line segments (i.e., a polygon), we can divide the region into triangles, and find the areas of each, and add them up. We can also find the area of regions bounded by arcs of circles, but the methods used are special and cannot be extended to other curves.

What we must do in the general case is to develop a method which can be applied to the widest numbers of curves, yet will give the right answer for, say, a triangle. The area we are interested in is bounded on the "north" by the graph of $y = f(x)$, which we will suppose is above the x axis, on the "south" by the axis, and by the vertical lines $x = a$ and $x = b$ on the "west" and "east". It looks something like Figure 10.11.

The basic plan to find the area is this: we will divide the interval $[a, b]$ into n equal subintervals, approximate the area over each subinterval by a rectangle, and then add up all these little areas to get an approximation of the total area, which we call A_n (for nth approximation). We then see what happens as the number of subintervals, n, gets

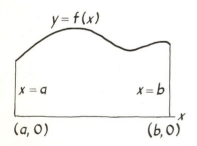

Figure 10.11

larger and larger. If the numbers A_n close in on a particular number, say A, we will write $\lim_{n \to \infty} A_n = A$ and actually *define* A to be the area in question.

We start then with the interval $[a, b]$ on the x axis. We pick an integer, n, and then divide the interval $[a, b]$ into n subintervals, labeling[1] the endpoints $x_0, x_1 \ldots x_{n-2}, x_{n-1}, x_n$, always being careful that a is labeled with x_0 and b is labeled with x_n. Since all of the subintervals are of the same length, we will call this length $\Delta_n x$ which is equal to $(b - a)/n$. Observe that as n gets large, $\Delta_n x$ gets close to zero.

Next, for each subinterval we approximate the area by a rectangle (see Figure 10.12). We do this by picking any point $\hat{x}_k$ between x_{k-1} and x_k, finding the value $f(\hat{x}_k)$ and using this as the height of our rectangle. The base is just $\Delta_n x$. This is the same thing as saying that the top of each rectangle must intersect the curve at some point. (One of the interesting things about this process is that it makes no difference in the final result where in $[x_{k-1}, x_k]$ we pick $\hat{x}_k$.)

Figure 10.12

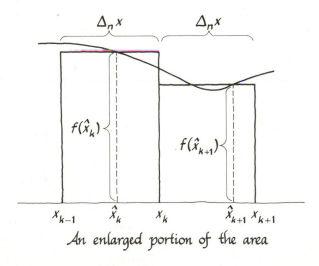

An enlarged portion of the area

---

[1]The little numbers below and to the right of the x's are called subscripts. They are used to keep track of a long list of (usually related) variables. They should be treated as if each was a different letter, x, y, z, etc., but they are more convenient to use, as we will see.

Then the area of our single approximating rectangle is $f(\hat{x}_k)(\Delta_n x)$. We do this for each of the subintervals, add up all of the areas of our approximating rectangles, and we will have that our nth approximation

$$A_n = f(\hat{x}_1)\Delta_n x + f(\hat{x}_2)\Delta_n x + \ldots + f(\hat{x}_n)\Delta_n x$$

Figure 10.13 shows this procedure for $n = 2$, 4, and 16. Notice how the difference between the actual area and the approximating area gets smaller as n gets larger.

Figure 10.13

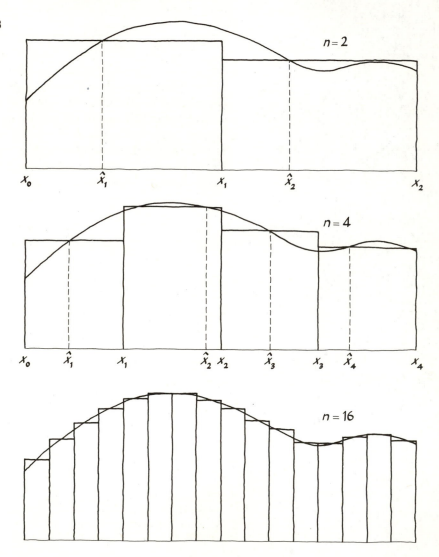

CHAPTER 10 / CALCULUS: A SAMPLER

Figure 10.14

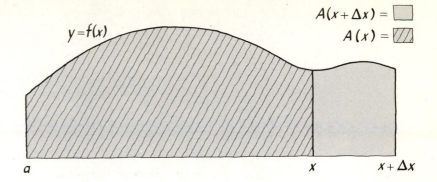

$$A(x + \Delta x) = \boxed{}$$
$$A(x) = \boxed{/\!/\!/}$$

$y = f(x)$

a x $x + \Delta x$

The principle used here is something like the following: Suppose we wish to construct a fence whose top must conform to a certain curve, but we are only allowed to use rectangular pieces of wood for the slats. Then the narrower we can pick the slats, the closer the top of the fence will match the required curve, even though each component piece has a flat top.

The actual evaluation of $\lim_{n \to \infty} A_n$ directly from the expression $A_n = (f(\hat{x}_1) + \ldots + f(\hat{x}_n))\Delta_n x$ is difficult, but we are aided by the following intuitive considerations.

Suppose we hold the left boundary, $x = a$, fixed, and think about what happens as we vary the righthand boundary from x to $(x + \Delta x)$. Then the area will depend on where this right boundary is, and we write down this dependence as $A(x)$.

Let's see if $A(x)$ has a derivative. The small portion to the right in Figure 10.14 represents $A(x + \Delta x) - A(x)$. We enlarge that portion in Figure 10.15, and observe that it is equal to the area of the rectangle whose base is Δx and whose altitude is $f(\tilde{x})$, where $\tilde{x}$ is *some point between x and $(x + \Delta x)$*. That is, $A(x + \Delta x) - A(x) = f(\tilde{x})(\Delta x)$.

Figure 10.15

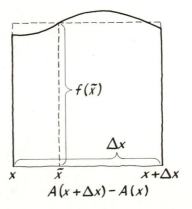

$f(\tilde{x})$

Δx

x $\tilde{x}$ $x + \Delta x$

$A(x + \Delta x) - A(x)$

Hence, if $\Delta x \neq 0$,

$$\frac{A(x + \Delta x) - A(x)}{\Delta x} = \frac{f(\tilde{x})(\Delta x)}{\Delta x} = f(\tilde{x})$$

Now as Δx nears zero, $\tilde{x}$ which is squeezed between x and $(x + \Delta x)$ has no place it can go except to close in on x, so that[1]

$$\lim_{\Delta x \to 0} \frac{A(x + \Delta x) - A(x)}{\Delta x} = f(x)$$

That is, $D_x A(x) = f(x)$, where $f(x)$ is the expression whose graph is the boundary curve at the top, or $A(x)$ is an antiderivative of $f(x)$.

Now suppose $F(x)$ is any other antiderivative of f. Since $F(x)$ is in the family we have called the general antiderivative of f, $F(x) + C = A(x)$. In particular,

$$F(b) + C = A(b)$$

which is the area under the graph of $f(x)$ from a to b that interests us. On the other hand, $A(a) = 0$; this can be seen geometrically (the area under the graph $f(x)$ from $x = a$ to $x = a$, that is, no area) or from the approximating sum

$$A_n = (f(\tilde{x}_1) + \ldots + f(\tilde{x}_n))\Delta_n x$$

by noting that, in this case,

$$\Delta_n x = \frac{a - a}{n} = 0$$

Hence

$$F(a) + C = A(a) = 0$$

or

$$C = -F(a)$$

so finally we have that the area we are interested in, the area under the graph of $f(x)$ from $x = a$ to $x = b$ and denoted by $A(b)$, is

$$A(b) = F(b) + C = F(b) - F(a),$$

where $F(x)$ is *any* antiderivative of $f(x)$. We will denote this number, $F(b) - F(a)$, by

$$\int_a^b f(x)\,dx$$

---

[1]We are assuming that f is nice enough so that if $\tilde{x}$ nears x, $f(\tilde{x})$ nears $f(x)$. For all the expressions we are interested in, and even for a much wider class, this is true.

and it is called *the definite integral of f(x) from a to b*. Once we know $F(x)$, we may also write it as

$$F(x)\big|_a^b$$

For example, the area under $y = x^3 + x$ from $x = 1$ to $x = 3$ is

$$\int_1^3 (x^3 + x)\,dx = \frac{x^4}{4} + \frac{x^2}{2}\bigg|_1^3$$

$$= \left(\frac{81}{4} + \frac{9}{2}\right) - \left[\frac{1}{4} + \frac{1}{2}\right]$$

$$= \frac{99}{4} - \frac{3}{4} = \frac{96}{4} = 24$$

In the above calculation, $F(x) = x^4/4 + x^2/2$, $F(b) = F(3)$ is in round brackets and $F(a) = F(1)$ is in square brackets.

This limiting procedure is quite general; observe that it did not depend on the fact that the graph of $f(x)$ was above the x axis. Any time we have a series of approximating sums like A_n, we can ask ourselves "Does $\lim_{n \to \infty} A_n$ exist?" If it does, then we call the limit the definite integral, and denote it by

$$\int_a^b f(x)\,dx$$

where $a = x_0$ and $b = x_n$.

As an illustration, suppose we wish to extend the concept of *arithmetic mean* to an algebraic expression over some interval. We know that if we have n numbers, $y_1, y_2, \ldots, y_n$, then their mean is $(1/n)(y_1 + y_2 + \ldots + y_n)$. But there are infinitely many values assumed by an algebraic expression over an interval. The basic plan is very much the same as before; we will find an nth approximation to the mean, M_n, and then see if $\lim_{n \to \infty} M_n$ has a limit, say M. If so, we will call M the mean of the expression. With a little adjustment, we can use the technique of antidifferentiation to evaluate the limit.

We start, as before, by dividing our interval, say $[a, b]$, into n equal subintervals, like this

We find the n values $f(x_1), f(x_2), \ldots, f(x_n)$, and then find their mean, which we will use as the nth approximating mean:

$$M_n = \frac{1}{n}\left(f(x_1) + f(x_2) + \ldots + f(x_n)\right)$$

If we were to investigate $\lim_{n \to \infty} M_n$ at this stage, we would be faced with a difficult problem, but things become easier if we multiply both sides by $(b - a)$; then

$$(b - a)M_n = \frac{b - a}{n} (f(x_1) + f(x_2) + \ldots + f(x_n))$$

or

$$(b - a)M_n = \Delta_n x (f(x_1) + f(x_2) + \ldots + f(x_n)),$$

since we have defined $\Delta_n x = (b - a)/n$. Observe that this is precisely like the approximating sum we had for area, so that

$$\lim (b - a)M_n = \int_a^b f(x)dx \qquad \text{or} \qquad \lim_{n \to \infty} M_n = \frac{1}{b - a} \int_a^b f(x)dx$$

For example, the work done in extending the end of a spring from its normal position to a distance of d units is kx^2 (Hooke's law), where k depends on the strength of the spring and the units used. To find the mean amount of work done in extending a spring six inches after it has already been pulled out two inches, we note that $a = 2$, $b = 6$ and $f(x) = kx^2$. Thus

$$M = \frac{1}{(6 - 2)} \int_2^6 kx^2 \, dx$$

$$= 1/4 \int_2^6 kx^2 = k/4 \, (x^3/3)|_2^6$$

$$= k/12 \, (216 - 27) = 189k/12 = 63k/4 \quad \text{units.}$$

There are many other uses for the definite integral besides mean values and areas under curves. For example, it is not too hard to show that the work done in moving an object from a to b is $\int_a^b f(x) \, dx$ if $f(x)$ is the moving force required at each point x in $[a, b]$. We can also use it to compute certain special volumes, such as the volume of a sphere or a cone, and, by an extension into three dimensions, volumes of a quite general nature. We can also combine it with the derivative to compute length along a curve, a surprisingly difficult problem, as it turns out.

There are also wide applications in the social sciences. In fact, the definite integral is at the heart of both probability and statistics. Also little can be done in some areas of economics without it. Who knows, perhaps history or English will be next? It would appear that the applications of this mathematical abstraction to real problems are practically endless.

EXERCISES 10.8

1. On four large sheets of paper, draw the graph of $y = x^2$ as accurately as possible from $x = 0$ to $x = 1$ so that the interval $[0, 1]$ is $6''$ long.
   (a) Divide the interval into two parts, $x_0 = 0$, $x_1 = \frac{1}{2}$ at $3''$, $x_2 = 1$ at $6''$. Estimate the area under $y = x^2$ between $x = 0$ and $x = 1$ by taking the first rectangle to have height $0^2 = 0$ and the second to have height $(\frac{1}{2})^2 = \frac{1}{4}$. Then estimate the same area by taking the first rectangle to have height $(\frac{1}{2})^2 = \frac{1}{4}$ and the second to have height $1^2 = 1$. (Note that the first estimate is equivalent to taking $\hat{x}_1 = x_0$ and $\hat{x}_2 = x_1$ while the second is equivalent to taking $\hat{x}_1 = x_1$ and $\hat{x}_2 = x_2$.) Calculate the difference between the two estimates as carefully as possible by measuring. Note that the first estimate is the worst possible *underestimate* and the second is the worst possible *overestimate*.
   (b) Divide $[0, 1]$ into four subintervals $x_0 = 0$ at 0, $x_1 = \frac{1}{4}$ at $1\frac{1}{2}''$, $x_2 = \frac{1}{2}$ at $3''$, $x_3 = \frac{3}{4}$ at $4\frac{1}{2}''$, and $x_4 = 1$ at $6''$. Make the worst possible underestimate by taking $\tilde{x}_1 = x_0$, $\tilde{x}_2 = x_1$, $\tilde{x}_3 = x_2$, and $\tilde{x}_4 = x_3$, and the worst possible overestimate by taking $\tilde{x}_1 = x_1$, $\tilde{x}_2 = x_2$, $\tilde{x}_3 = x_3$, and $\tilde{x}_4 = x_4$. Find the difference, and compare with your answer in (a).
   (c) Repeat the processes for $n = 8$ and $n = 16$.
   (d) What conclusion can you come to about the sizes of the under- and overestimates of the area?
   (e) How does this convince you that $\lim_{n \to \infty} A_n$ exists?

2. Find the area under the horizontal line $y = 6$ between $x = 2$ and $x = 7$, using the definite integral. What does this tend to confirm? Try to generalize the result yourself.

3. Prove that the area computed by methods of calculus agrees with the usual definition of area for a right triangle, by finding the area under the line given by $y = (h/b)x$ between $x = 0$ and $x = b$. Is it now necessary to prove this for any triangle? Explain.

4. Find the area bounded by $x = 1$, $x = 4$, the graph of $y = 4x^3 - 6x + 3$, and the x axis.

5. If $f(x) \geq g(x)$ for all x, we define the area bounded by the graph of $f(x)$, the graph of $g(x)$, $x = a$, and $x = b$, to be $\int_a^b (f(x) - g(x))dx$.
   (a) Find the area bounded by the graphs of $y = x^3$ and $y = x^2$ for x between $x = 0$ and $x = 1$; also between $x = 1$ and $x = 2$, and between $x = 0$ and $x = 2$. (*Hint*: Make careful drawings.)

(b) Find the area bounded by the x axis and $y = x^3$ between $x = -2$ and $x = 1$. (*Hint*: See the previous hint.)

6. Water is flowing from an irregularly shaped hole in a dam at the rate of $10{,}000 - t^2$ gallons per second t seconds after the break occurred.
   (a) How long before the water stops?
   (b) What is the average rate of flow between $t = 10$ and $t = 20$, between $t = 60$ and $t = 70$? between $t = 100$ and $t = 110$?

*7. Using the fact that work done is $\int_a^b f(x)dx$, as described in the text, and that the force required to extend a spring is proportional to the distance d it is extended beyond rest, to develop Hooke's Law, $W = kd^2$. Here k is some constant which depends upon the strength of the spring and the units used. What assumption is made about d relative to the spring?

REFERENCES

Hofmann, Joseph E. *Classical Mathematics*. Translated by H. O. Midonich. Philosophical Library, Inc., New York, 1959.

Hogben, Lancelot. *Mathematics for the Millions*, 4th ed. W. W. Norton & Company, Inc. New York, 1968.

Hooper, Alfred. *Makers of Mathematics*. Random House, Inc., New York, 1948.

Kline, Morris. *Mathematics, A Cultural Approach*. Addison-Wesley Publishing Co., Inc., Reading, Mass., 1962.

Kline, Morris. *Mathematics and the Physical World*. Doubleday Publishing Company, Anchor Books, Garden City, N.Y., 1963.

Newman, James R. *The World of Mathematics*. 4 vols. Simon & Schuster, Inc., New York, 1956.

Peter, Rosza. *Playing with Infinity: Mathematics for Everyman*. Translated by Z. P. Dienes. Simon & Schuster, Inc., New York, 1964.

Bibliography

The following is a list of books for additional reading. Some are hard and some are easy. Except in rare cases I have not indicated which is which because I don't wish to frighten you away from the harder ones. In any case, there is no book on the list which cannot be read, at least in part, by some students in the course.

Not all of the books take the same point of view toward mathematics as we do in this course. Things would be pretty uninteresting if we all thought alike.

The numerals in parentheses indicate the chapters of this text which come closest to the subject matter of the individual book. Where no numerals are given, the book covers pretty much the entire course or is somewhat to one side altogether.

Aaboe, Asgar. *Episodes From the Early History of Mathematics.* Random House, New York, 1964. Fairly nontechnical; uses modern notation. (2, 4)

Abbott, Edwin A. *Flatland, A Romance in Many Dimensions.* Blackwell, Oxford, 1950.

Barker, Stephen F. *Philosophy of Mathematics.* Prentice-Hall, Englewood Cliffs, New Jersey, 1964. (1)

Beckmann, Petr. *A History of π.* 2d edition. Golem Press, Boulder, Colorado, 1971. (2, 4)

Bell, Eric Temple. *The Development of Mathematics.* McGraw-Hill, New York, 1945.

———. *The Handmaiden of the Sciences.* Reynal and Hitchcock, 1937. Mathematics and Applications. (1)

———. *Men of Mathematics.* Simon & Schuster, New York, 1937, 1965.

Bell, Arthur E. *Newtonian Science.* Arnold, London, 1960. (9)

Boyer, Carl. *The History of Calculus.* Dover, New York, 1959. For the seriously interested student.

Brewster, David. *Memoirs of . . . Isaac Newton.* 2 vols. Johnson Reprint, New York, 1965. (9)

Brodetsky, S. *Sir Isaac Newton.* Luce, Boston, 1927. (9)

Broehm, George A. W., and the editors of Fortune. *The New World of Math.* Dial Press, New York, 1959. (1)

Brydegaard, M., and H. E. Inskeep, eds. *Reading for Geometry from the Arithmetic Teacher.* National Council of Teachers of Mathematics, Washington, D.C., 1970. (4)

Cardano, Gerolamo. *The Great Art or the Rules of Algebra.*

Translated and edited by T. Richard Witmer. M.I.T. Press, Cambridge, Massachusetts, 1968. Heavily annotated, basically from the 1545 edition of *Artis Magnae.* (3)

Cajori, Florian. *A History of Elementary Mathematics.* Macmillan, New York, 1924. (2, 3)

———. *A History of Mathematical Notations.* 2 vols. Open Court Publishing, La Salle, Illinois, 1928. (1, 2, 3, 4)

Clifford, William Kingdon. *The Common Sense of the Exact Sciences.* Knopf, New York, 1946.

Coolidge, Julian L. *The Mathematics of Great Amateurs.* Dover, New York, 1963 (reprint of Oxford University Press, 1949).

Copernicus. *On the Revolutions of the Heavenly Spheres.*

Translated by C. G. Waths. Great Books of the Western World, Vol. 16. Encyclopedia Britannica, London, 1952. Not easy (9)

Cundy, H. Martyn, and A. P. Rollett. *Mathematical Models.* 2d edition. Oxford University Press, London. 1961. (4)

Dannington, G. Waldo. *Carl Friedrich Gauss, Titan of Science.* Hafner, New York, 1955.

Dantzig, Tobias. *The Bequest of the Greeks.* Scribners, New York, 1955. Popular, easy to read.

——. *Number, the Language of Science.* Macmillan, New York, 1954. (2)

D'Elia, Pasquale M. S. J. *Galileo in China.* Translated by R. Suter and M. Sciascia. Harvard University Press, Cambridge, Massachusetts, 1960.

De Santillana, Giorgio. *The Crime of Galileo.* University of Chicago Press, Chicago, 1955. (9)

Descartes, Rene. *The Geometry of Rene Descartes.* Translated by D. E. Smith and M. L. Latham. Dover, New York, 1954. Translation of *La Geometrie.*

Drake, Stillman. *Discoveries and Opinions of Galileo.* (translated and annotated). Doubleday, Anchor, Garden City, New York, 1957. (9)

——. *Galileo Studies, Personality, Tradition, and Revolution.* University of Michigan Press, Ann Arbor, 1970. A series of essays. (9)

Dubnov, Ya S. *Mistakes in Geometric Proofs.* Translated

by A. Henn and O. A. Titetbaum. Heath, Boston, 1963. (1, 4)

Easley, J. A. Jr., and Maurice M. Tatsuoka. *Scientific Thought, Cases from Classical Physics.* Allyn & Bacon, Boston, 1968. Chapter 2 on linear motion. (9)

Fang, J. *Bourbaki.* Paidelia Press, Hauppauge, New York, 1970.

Galilei, Galileo. *Dialogue Concerning the Two Chief World Systems.* Translated by Stillman Drake. University of California Press, 1967.

——. *Dialogue Concerning the Two New Sciences.* Translated by H. Crew and A. de Salvio. Great Books of the Western World, Vol. 28. Encyclopedia Britannica, London, 1952.

——. *Dialogue on the Great World System.* (Salusbury translation.) Chicago University Press, Chicago, 1953.

Gamow, George. *One, Two, Three . . . Infinity.* Parts I and II. Revised Edition. Viking, New York, 1961. (2, 4)

Gow, James. *A Short History of Greek Mathematics.* Chelsea, 1968. (Revised reprint of original, Cambridge, England, 1884.) (2, 4)

Hadamard, Jacques. *The Psychology of Invention in the Mathematical Field.* Dover, New York, 1954; also Princeton University Press, 1945.

Hankins, Thomas L. *Jean d'Alembert, Science and the*

Enlightenment. Oxford University Press, London, 1970.

Heath, Thomas. *A History of Greek Mathematics.* Vols. I, II. Oxford University Press, London, 1960. An exhaustive work but useful for the interested student for selected readings. (2, 4)

——. *A Manual of Greek Mathematics.* Dover, New York, 1963 (reprint of Oxford University Press, 1931). For the general reader, a reworking of *A History of Greek Mathematics.* (2, 4)

——. *The Thirteen Books of Euclid's Elements.* Vols. I, II, III. Dover, New York, 1956. (4)

Hofmann, Joseph E. *Classical Mathematics.* Translated by H. O. Midonich. Philosophical Library, New York, 1959. (10)

——. *The History of Mathematics.* Translated by F. Gaynor and H. O. Midonich. Philosophical Library, New York, 1957. (2, 3)

Hogben, Lancelot. *Mathematics for the Millions.* 4th edition. W. W. Norton, New York, 1968. Historical emphasis, especially recommended for the student who wants to learn more mathematics as painlessly as possible. (1, 2, 3, 4, 10)

——. *Mathematics in the Making.* Doubleday, New York, 1960.

Hooper, Alfred. *Makers of Mathematics.* Random House, New York, 1948. (3, 4, 10)

Huff, Durrell. *How to Lie With Statistics.* Norton, New York, 1954. (6)

Huntley, H. E. *The Divine Proportion: A Study in Mathematical Beauty.* Dover, New York, 1970. (2, 4)

BIBLIOGRAPHY

Kasner, E., and J. Newman. *Mathematics and the Imagination*. Simon & Schuster, New York, 1943.

Kemeny, John G. *Random Essays on Mathematics, Education and Computers*. Prentice-Hall, Englewood Cliffs, New Jersey, 1964.

Kesten, Hermann. *Copernicus and His World*. Roy, New York, 1945.

Kline, Morris. *Mathematics and the Physical World*. Doubleday, Anchor, New York, 1963; also Thomas Y. Crowell, New York, 1959. (1, 2, 3, 9, 10)

——. *Mathematics, A Cultural Approach*. Addison-Wesley, Reading, Massachusetts, 1962. (1, 4, 9, 10)

Kline, Morris. ed. *Mathematics in the Modern World*. W. H. Freeman, San Francisco, 1969. (4, 6)

——. *Mathematics in Western Culture*. Oxford University Press, New York, 1953. Designed for the nonscientist. Discusses the arithmetization of our society in some detail. (1)

Kramer, Edna E. *The Main Stream of Mathematics*. Oxford University Press, New York, 1955.

Labban, Kushyar Ibn. *Principles of Hindu Reckoning*. Translated by M. Levey and M. Petruch. University of Wisconsin Press, Madison, 1965. (2)

Laplace, Pierre Simon. *A Philosophical Essay on Probabilities*. Translated by F. W. Truscott and F. L.

Emory. Dover, New York, 1951. Nontechnical. (6)

Loomis, Elisha Scott. *The Pythagorean Proposition*. National Council of Teachers of Mathematics, Washington, D.C., 1968. (4, 7)

Mach, Ernst. *The Science of Mechanics*. Translated by T. J. McCormack. Open Court Publishing, London, 1942. (9)

Manuel, Frank E. *A Portrait of Isaac Newton*. Harvard University Press, Cambridge, 1968. (9)

Mathematical Thinking in the Behavioral Sciences. Readings from the *Scientific American*. Freeman, San Francisco, 1968.

Maxwell, J. Clark. *Matter and Motion*. Dover, New York, 1877. (9)

McKay, Herbert. *The World of Numbers*. Macmillan, New York, 1946. (1)

McMullin, Ernan, ed. *Galileo, Man of Science*. Basic Books, New York, 1967. (9)

Meschkowski, Herbert. *Ways of Thought of Great Mathematicians*. Translated by J. Dyer-Bennet. Holden-Day, San Francisco, 1964.

COSRIMS, *The Mathematical Sciences*. M.I.T. Press, Cambridge, Massachusetts, 1969. (1)

Namer, Emile. *Galileo, Searcher of the Heavens*. Translated by S. Harris. McBride, New York, 1931. Astronomical emphasis. (9)

Neugebauer, O. *The Exact Science in Antiquity*. 2nd edition. Brown University Press, Providence, 1957. (4)

Newman, James R. *The World of Mathematics*. 4 volumes. Simon & Schuster, New York, 1956. (1, 2, 3, 4, 9, 10)

Newsom, Carroll V. *Mathematical Discourses, The Heart of Mathematics*. Prentice-Hall, Englewood Cliffs, N.J., 1964. (1)

Newton, Isaac. *Mathematical Principles of Natural Philosophy*. Translated by A. Motte; revised by F. Cajori. Great Books of the Western World, Vol. 34. Encyclopedia Britannica, London, 1952. Tough going. (9)

North, J. D. *Isaac Newton*. Oxford University Press, London, 1967. (9)

Peter, Rozsa. *Playing With Infinity: Mathematics for Everyman*. Translated by Z. P. Dienes. Simon & Schuster, New York, 1962. (2, 5, 8, 10)

Polya, G. *How to Solve It*. Doubleday, Anchor, Garden City, New York, 1957; also Princeton University Press, 1945. (1)

Ptolemy. *The Almagest*. Translated by R. Catesby Taliaferro. Great Books of the Western World, Vol. 16. Encyclopedia Britannica, London, 1952. Astronomy; not easy. (9)

Pullan, J. M. *The History of the Abacus*. Praeger, New York, 1968. (2)

Reid, Constance. *From Zero to Infinity*. Thomas Y. Crowell, New York, 1960. (2)

——. *Hilbert*. Springer-Verlag, New York, 1970.

Renyi, Alfred. *Dialogues on Mathematics.* Holden-Day, San Francisco, 1967. (1, 9)

Russell, Bertrand. *The Impact of Science on Society.* Columbia University Press, New York, 1953. (1)

Sanford, Vera. *A Short History of Mathematics.* Houghton Mifflin, Boston, 1958.

Sarton, George. *The Study of the History of Mathematics and The Study of the History of Science.* 2 vols. bound as one. Dover, New York, 1957.

Scott, J. F. *A History of Mathematics.* Barnes & Noble, New York, 1969.

Seeger, Raymond J. *Galileo Galilei, His Life and His Works.* Pergamon Press, New York, 1966. (9)

Shamos, H. Morris, ed. *Great Experiments in Physics.* Holt, Rinehart and Winston, New York, 1959.

Struik, Dirk. *A Concise History of Mathematics.* Dover, New York, 1967.

——. ed. *A Source Book in Mathematics, 1200-1800.* Harvard University Press, Cambridge, 1969.

Sutton, O. G. *Mathematics In Action.* Bell, London, 1958. (1, 9)

Taylor, R. Emmett. *No Royal Road, Luca Pacioli and His Times.* University of North Carolina Press, Durham, 1942.

Van Der Waerden. *Science Awakening.* (A. Dresden, trans.) Noordhoff, Groningen, Holland, 1954.

Weaver, Warren. *Lady Luck, The Theory of Probability.* Doubleday, Anchor, Garden City, New York, 1963. (6)

Webster. *Topology, The Rubber-Sheet Geometry.* St. Louis, 1960.

Wiener, Norbert. *Ex-Prodigy, My Childhood and Growth.* Simon & Schuster, New York, 1953.

——. *I Am a Mathematician.* Doubleday, New York, 1956.

Whitehead, Alfred North. *The Interpretation of Science.* Edited by A. H. Johnson. Bobbs-Merrill, New York, 1961. (1, 9)

——. *An Introduction to Mathematics.* Oxford University Press, New York, 1958. (1, 3, 9)

Willerding, Margaret. *Mathematical Concepts, A Historical Approach.* Prindle, Weber & Schmidt, Boston, 1967. (2)

Zippin, Leo. *Uses of Infinity.* Random House, New York, 1962. (4, 7)

Credits

Graphic motif for cover design by Conrad Guertin

Courtesy Cluny Museum, Time-LIFE, Pierre Boulat, *Madame Arithmetic*, part opening photo, opposite **p. 1**

From Bill Finch, **p. 3**

From Scala, New York/Florence, The Vatican, *The School of Athens* by Raphael, part opening photo, **p. 26**

From Barbara Marshall, **p. 29**

From David Wunsch, **p. 52**

From Manfred Kage from Peter Arnold, **p. 73**

From The Library of Istanbul University, Derek de Solla Price, a 16th Century Persian illumination, part opening photo, **p. 90**

From Erik Hansen, **p. 93**

From Frank Siteman/Stock, Boston, **p. 118**

From Frank Siteman/Stock, Boston, **p. 143**

From Talbot Lovering/Stock, Boston, **p. 176**

From Frank Siteman/Stock, Boston, **p. 208**

From Bill Finch, **p. 229**

Illustrations by Jane Wentzell

Index

Populations, definition of, 135
Positional notation, 30
  for fractions, 42
Position function, 231
Postulates, 17
  Euclidean, 75, 76–80
Primitive events, 122–123
Probabilities
  assignment of, 120–121, 122–123,
  125–129
  conditional, 131–133
  definition of, 120
  of sample space, 121–122
  statistics and, 139–141
Problem, translation into
  mathematics, 19–21
Proof
  by contradiction, 107–108
  definition of, 17
  Euclid's fifth postulate, attempts
    to, 78
  synthetic, 143
Pseudosphere, 80
Pythagorean theorem, 79

Quadratic formula, 68

Random sampling, definition of, 139
Range of data, 136
  of function, 169–170
Rational numbers, 40–41
  addition of, 41
  decimal expansion of, 42–45
  location on number line, 145–146
  lowest terms of, 47
Reasoning, 7–8
  deductive, 8–14, 17
  inductive, 8, 22–23
Reductio ad absurdum, 107–108
Relation, function compared with, 170
Renieri, Vincenzio, 214
Right angle, size of, 76
Russell's paradox, 116–117

Sample
  definition of, 135

random, 139
  size of, 140–141
Sample space
  construction of, 121–122
  definition of, 120
  probability of, 122
Secant line, slope of, 235–236
Sentences, algebraic, 59–61
Sentential calculus, 93
Sets, 108–110
  empty, 111
  events as, 120
  limitations on, 116–117
  properties of, 110–115
  universal, 114
Sine, 86
  determination of, 164–167
Slope, 157
Speed, definition of, 220
Square roots
  addition of, 48–49
  multiplication of, 48
  sign of, 48
Squares, of odd and even integers, 47
Standard deviation, 136–137, 138
  calculation of, 137
Standard position of an angle, 164
Statements
  descriptive, 109–110
  equal, 100
  list as, 109
  in probability, 120
  in sentential calculus, 94
Statistics, 134–138
  probability and, 139–141
Subevent, definition of, 125
Subscripts, 31, 261n
Subset, definition of, 111
Subtraction
  in base 8, 31–33
  of integers, 37
Syllogism, 8–13
Synthetic geometry, 143

Tangent, 87–88
  defined, 87
  lacking, 255–256
  line, defined, 234–238

Theorems, 17
  for absolute value, 195–197
  for differentiation, 247–249
  discovery of, 22
  of inequality, 179–182
  Pythagorean, 79
Transcendentals, 49
Triangle inequality, 197
Triangles, similar, 86
Trichotomy property, 179
Trigonometry, 86–88
  analytic, 163–168
Truth, problem of defining, 14–15
Truth tables, 96
  construction of, 97–98
  equal statements and, 100

Undefined notions, 18
Union, of sets, 112
Unit, number line and, 144
Unit axioms, 55–56
Unit circle, 164
Universal set, 114
  in probability, 120

Validity, 10–14
Value, absolute, 194–197
Velocity
  average, 230–231
  instantaneous, 234

Weight, definition of, 217
Whole numbers, *see* Counting numbers

x axis, 148

y axis, 148

Zero
  division by, 39–40
  special properties of, 35

INDEX